Identifying and Supporting Productive STEM Programs in Out-of-School Settings

Committee on Successful Out-of-School STEM Learning

Board on Science Education

Division of Behavioral and Social Sciences and Education

NATIONAL RESEARCH COUNCIL
OF THE NATIONAL ACADEMIES

THE NATIONAL ACADEMIES PRESS
Washington, D.C.
www.nap.edu

THE NATIONAL ACADEMIES PRESS • 500 Fifth Street, N.W. • Washington, DC 20001

NOTICE: The project that is the subject of this report was approved by the Governing Board of the National Research Council, whose members are drawn from the councils of the National Academy of Sciences, the National Academy of Engineering, and the Institute of Medicine. The members of the committee responsible for the report were chosen for their special competences and with regard for appropriate balance.

This study was supported by Contract No. DRL-1339083 between the National Academy of Sciences and the National Science Foundation. Any opinions, findings, conclusions, or recommendations expressed in this publication are those of the author(s) and do not necessarily reflect the views of the organizations or agencies that provided support for the project.

International Standard Book Number-13: 978-0-309-37362-3
International Standard Book Number-10: 0-309-37362-X
Library of Congress Conrol Number: 2015945523

Additional copies of this report are available from National Academies Press, 500 Fifth Street, NW, Keck 360, Washington, DC 20001; (800) 624-6242 or (202) 334-3313; http://www.nap.edu.

Printed in the United States of America

Suggested citation: National Research Council. (2015). *Identifying and Supporting Productive STEM Programs in Out-of-School Settings.* Committee on Successful Out-of-School STEM Learning. Board on Science Education, Division of Behavioral and Social Sciences and Education. Washington, DC: The National Academies Press.

THE NATIONAL ACADEMIES

Advisers to the Nation on Science, Engineering, and Medicine

The National Academy of Sciences is a private, nonprofit, self-perpetuating society of distinguished scholars engaged in scientific and engineering research, dedicated to the furtherance of science and technology and to their use for the general welfare. Upon the authority of the charter granted to it by the Congress in 1863, the Academy has a mandate that requires it to advise the federal government on scientific and technical matters. Dr. Ralph J. Cicerone is president of the National Academy of Sciences.

The National Academy of Engineering was established in 1964, under the charter of the National Academy of Sciences, as a parallel organization of outstanding engineers. It is autonomous in its administration and in the selection of its members, sharing with the National Academy of Sciences the responsibility for advising the federal government. The National Academy of Engineering also sponsors engineering programs aimed at meeting national needs, encourages education and research, and recognizes the superior achievements of engineers. Dr. C. D. Mote, Jr., is president of the National Academy of Engineering.

The Institute of Medicine was established in 1970 by the National Academy of Sciences to secure the services of eminent members of appropriate professions in the examination of policy matters pertaining to the health of the public. The Institute acts under the responsibility given to the National Academy of Sciences by its congressional charter to be an adviser to the federal government and, upon its own initiative, to identify issues of medical care, research, and education. Dr. Victor J. Dzau is president of the Institute of Medicine.

The National Research Council was organized by the National Academy of Sciences in 1916 to associate the broad community of science and technology with the Academy's purposes of furthering knowledge and advising the federal government. Functioning in accordance with general policies determined by the Academy, the Council has become the principal operating agency of both the National Academy of Sciences and the National Academy of Engineering in providing services to the government, the public, and the scientific and engineering communities. The Council is administered jointly by both Academies and the Institute of Medicine. Dr. Ralph J. Cicerone and Dr. C. D. Mote, Jr., are chair and vice chair, respectively, of the National Research Council.

www.national-academies.org

COMMITTEE ON SUCCESSFUL OUT-OF-SCHOOL STEM LEARNING

ERIC JOLLY (*Chair*), Science Museum of Minnesota, St. Paul

BRONWYN BEVAN, Exploratorium Institute for Research and Learning, San Francisco

JANE BUIKSTRA, Center for Bioarchaeological Research, Arizona State University

JACQUELYNNE ECCLES, Center for Teaching Excellence, University of California, Irvine

JOHN FALK, College of Education, Oregon State University and Institute for Learning Innovation, Corvallis

MAYA GARCIA, Office of the State Superintendent of Education, Government of the District of Columbia

LESLIE GOODYEAR, Education Development Center, Inc., Waltham, MA

LYNN S. LIBEN, Department of Psychology, Pennsylvania State University

MILBREY MCLAUGHLIN, Graduate School of Education, Stanford University

VERA MICHALCHIK, Office of the Vice Provost for Teaching and Learning, Stanford University

NANCY PETER, Out-of-School Time Resource Center, University of Pennsylvania

CARY SNEIDER, Center for Education, Portland State University

JILL WALAHOSKI, State 4-H Department, University of Nebraska

MICHAEL FEDER, *Study Director*

JOANNA ROBERTS, *Program Assistant*

ARGENTA PRICE, *Christine Mirzayan Science and Technology Fellow*

HEIDI SCHWEINGRUBER, *Director,* Board on Science Education

MARTIN STORKSDIECK, *Director,* Board on Science Education (until June 2014)

CONTENTS

EXECUTIVE SUMMARY

The ways in which young people learn about science, technology, engineering, and mathematics (STEM) has fundamentally changed in the past decade. More so than ever, young people now have opportunities to learn STEM in a wide variety of settings, including clubs, summer programs, museums, parks, and online activities. They spend more time in supervised programs outside of school, and they have greater access to on-demand learning resources and opportunities. At the same time, STEM learning outside of school has become a focal piece of the education opportunities provided by many national nonprofit organizations, statewide education networks, federal programs, and corporate and family foundations. And there is growing evidence that opportunities to learn STEM outside of school directly affect what is possible inside classrooms, just as what happens in classrooms affects out-of-school learning.

The Committee on Successful Out-of-School STEM Learning was charged with outlining the criteria that policy makers, program developers, and other stakeholders can use to identify effective out-of-school STEM settings and programs. It was also charged with identifying those criteria for which data are readily available and those for which further work is needed to develop appropriate data sources. To address its charge, the committee organized a National Summit on Successful Out-of-School STEM Learning, reviewed relevant research, and commissioned papers to synthesize existing research.

Research and evaluation related to learning outside of school have been conducted by professionals from many fields, including youth development, cognitive and social development, informal learning, and out-of-school time. Evidence from these fields shows that STEM learning results from the dynamic interactions that occur over time among the diverse settings in which learning occurs (e.g., youth groups, hobby clubs, museums, libraries, schools, home), the community and culture in which they are embedded, and the characteristics of the learner (e.g., interests, dispositions, values). Within this dynamic system, out-of-school programs have been shown to:

- contribute to young people's interest in and understanding of STEM,
- connect young people to caring adults who serve as role models, and
- reduce the achievement gap between young people from low-income and high-income families.

Research and evaluation findings are not yet robust enough to determine which programs work best for whom and under what circumstances. The limitations of the existing research are due to the many types of out-of-school STEM programs and the difficulties of measuring the outcomes of such programs. The findings are strong enough, however, to identify three criteria of programs that produce positive outcomes for learners: they are engaging, responsive, and make connections.

Criteria for Identifying and Developing Productive STEM Out-of-School Programs

ENGAGING

Engage Young People Intellectually, Academically, Socially, and Emotionally

- Program provides first-hand experiences with phenomena and materials.
- Program engages young people in sustained STEM practices.
- Program establishes a supportive learning community.

RESPONSIVE

Respond to Young People's Interests, Experiences, and Cultural Practices

- Program positions STEM as socially meaningful and culturally relevant.
- Program supports young people to collaborate and to take on leadership roles in STEM learning activities.
- Program positions staff as co-investigators and learners alongside young people.

MAKE CONNECTIONS

Connect STEM Learning in Out-of-School, School, Home, and Other Settings

- Program connects learning experiences across settings.
- Program leverages community resources and partnerships.
- Program actively brokers additional STEM learning opportunities.

To better understand how productive out-of-school STEM programs contribute to young people's interest in and understanding of STEM, evaluations must address individual, program-, and community-level outcomes. Building the capacity to generate evidence at these three levels will lead to a clearer picture of how programs affect outcomes across settings and time. In addition, mapping the STEM learning assets of communities can inform decisions about where further investment is needed and support connections among STEM learning opportunities. Innovative measures can illustrate what programs work for whom and under what circumstances and can make it easier to compare and aggregate program outcomes.

As measurement work moves forward, it is important to avoid two common mistakes. One is to depend entirely on short-term student learning outcomes as indicators of productive programs rather than recognizing more complex and varied outcomes. The second is to measure outcomes in ways that alter the nature of productive programs or ignore the differences in out-of-school programs in order to generate comparative or aggregate data.

The committee identified six actions that policy makers, program developers, and stakeholders should take to develop and support productive programs:

- Understand the local conditions for community programs that support STEM learning: **Build a map and bridge the gaps.**
- Design programs to achieve access, equity, continuity, and coherence: **Connect young people to opportunities to learn.**
- Support the use of creative and responsive approaches to evaluate the success of programs at the individual, program, and community levels: **Support innovative evaluation approaches.**
- Increase the professionalization of out-of-school program leaders and staff: **Provide professional development.**
- Strengthen the STEM learning infrastructure: **Build an infrastructure that will last.**
- Invest in research to improve our understanding of STEM learning in out-of-school programs: **Explore how STEM learning ecosystems work.**

Some of these activities can only be undertaken at the local level; some will require national-level involvement. All of them need to be undertaken with sensitivity to the young people who have historically been underserved by STEM learning programs, including girls, ethnic minorities, and those from economically marginalized communities. Together, the actions above can support productive out-of-school STEM programs.

WHERE AND HOW YOUNG PEOPLE LEARN STEM

Over the past decade, there has been a fundamental change in the way that learning is organized and supported. As family work patterns shift, children and youths are spending more time in supervised educational programs before and after school, on weekends, and during summers and other holidays.[1] At the same time, more children and youths regularly access on-demand digital learning resources and opportunities, including online communities and resource collections. Thus, education can no longer be defined solely by what happens in a schoolroom. Indeed, a substantial body of research demonstrates that deep learning develops across multiple settings and timeframes.[2] What happens outside the classroom directly affects what is possible inside the classroom and vice versa.[3]

About the Report

This report, funded by the National Science Foundation, provides guidance for designing and implementing out-of-school science, technology, engineering, and mathematics (STEM) learning opportunities for all young people (ages 5-18). The intended audiences of the report are local, state, and federal policy makers, out-of-school STEM program developers, and both classroom educators and out-of-school educators. To address the statement of task for this study, the report describes the role that out-of-school programs play in deepening and broadening young people's

Statement of Task

An ad hoc committee will plan and conduct a public workshop to explore criteria for identifying highly successful practices in the area of STEM education in out-of-school settings, with a focus on designed settings and programs targeted at children and youth, through examination of a select set of examples. The committee will determine some initial criteria for nominating successful practices to be considered at the workshop. The examples included in the workshop must have been studied in enough detail to provide evidence to support claims of success. Discussions at the workshop will focus on refining criteria for success, exploring models of "best practice," and an analysis of factors that evidence indicates lead to success. The discussion from the workshop will be synthesized and combined with a literature review of peer-reviewed and gray-literature publications for a short, committee-authored consensus report that would outline criteria for identifying effective out-of-school STEM settings and programs and identify those criteria for which data are readily available and those where further work is needed to develop appropriate data sources.

access to multiple, high-quality STEM learning opportunities. Such programs are important in building a STEM-engaged and STEM-literate society and workforce. The report focuses on STEM learning that occurs in out-of-school programs that are designed and led by adults, and structured for youths.[a] Included are afterschool programs, summer and weekend classes, and apprenticeship opportunities.[b] We identify the features of productive out-of-school STEM programs, review the evidence of the effects of out-of-school STEM programs, discuss the capacity needs of program staff, and provide a framework for improving evaluations.

Our conclusions and recommendations are based on a review of the literature on out-of-school STEM learning programs and practices, and, more broadly, on STEM learning. We also hosted a national summit on out-of-school STEM programs and commissioned a set of research reviews to gather critical information for the report: see Appendix A for the summit agenda and Appendix B for the list of commissioned papers. Although the committee reviewed all the evidence on learning STEM in out-of-school programs that we could identify, this study does not include a detailed literature review of that work because it is beyond the scope of this study.

Thinking Systemically About STEM Learning

Over the past decade, many policy makers, funders, communities, and educators have come together to align resources to enrich what has been called the *STEM learning ecosystem.*[4] This phrase refers to the dynamic interaction among individual learners, diverse settings where learning occurs, and the community and culture in which they are embedded.[5] A STEM learning ecosystem[6] includes all of a community's STEM-rich assets, which include

- *designed settings,* such as schools, clubs, museums, and youth programs;
- *naturalistic settings*, such as city parks, waterways, and forests and deserts;
- *people and networks of people,* such as practicing STEM professionals, educators, enthusiasts, hobbyists, and business leaders who can serve as inspiration and role models; and
- *everyday encounters* with STEM, such as on the Internet, on television, on the playground, or during conversations with family members and other young people.[c]

In a STEM learning ecosystem, children are at the center of the model because children are influenced directly by other people (e.g., family, friends) and settings (e.g., schools, neighborhoods) and indirectly by their environment and culture. In turn, children themselves shape and influence

[a]Learning opportunities that take place outside of school have been referred to in many ways, including informal learning, nonformal learning, life-long learning, out-of-school time learning, and free-choice learning. We use the term "out-of-school programs" to focus on the particular set of learning opportunities in our charge.

[b]We exclude designed, unsupervised youth learning opportunities, such as television, radio, Internet, and social media projects.

[c]Everyday encounters are outside the focus of this report but are an important part of the STEM learning ecosystem.

the environment through their interests, dispositions, and values. Time is included in this model to illustrate that there are constant changes in children themselves and in the surrounding context. For example, the cognitive, emotional, social, and motivational qualities that young people bring to learning experiences are constantly evolving as they mature and accumulate experiences. Each learning experience has the potential to augment and be augmented by these qualities, leading to a dynamic interplay over time between the qualities of young people and those of learning environments.[7] Thus, from an ecosystems perspective of STEM learning, connections among learners, community assets, and the broader culture are critical for supporting young people's learning.

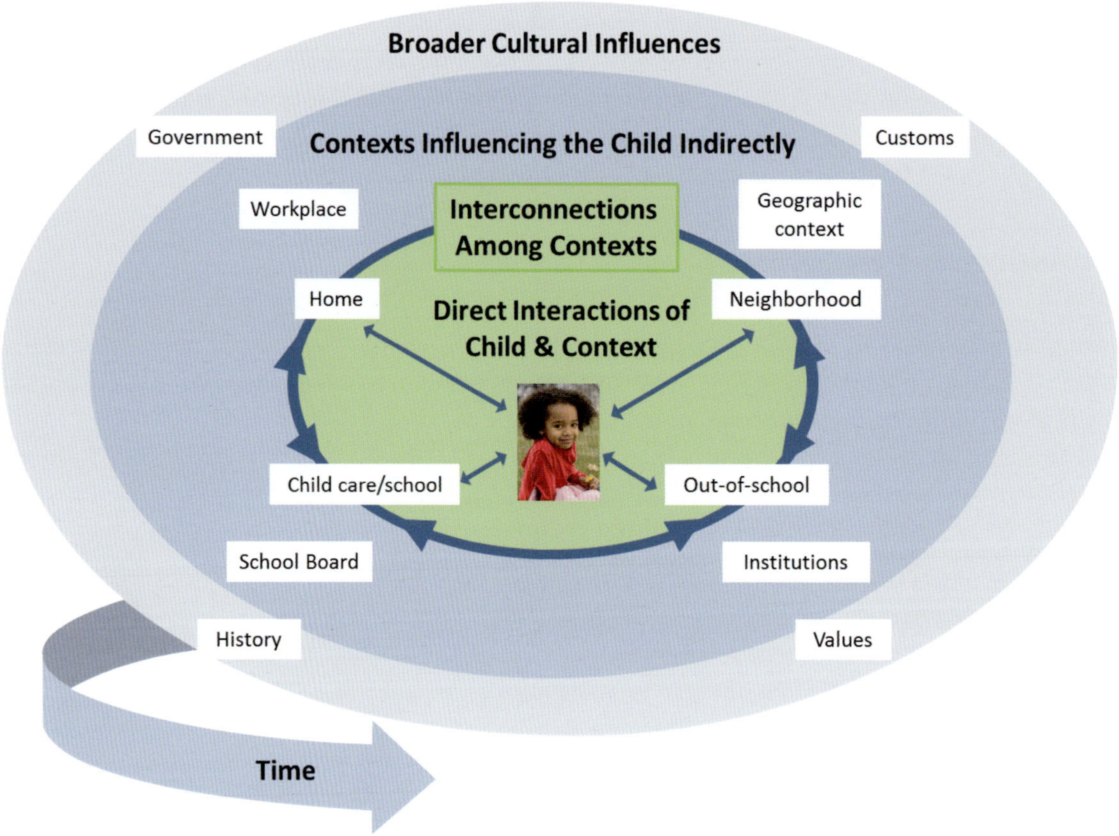

STEM learning ecosystem model.

NOTE: This representation, of the learning ecosystem model is based on Bronfenbrenner's ecological model of human development first published in 1977. The innermost circle represents interactions that directly involve both child and an embedding context (e.g., child ↔ school). The next level shows connections among the immediately embedding contexts themselves. These also affect the child's experiences (e.g., quality of family ↔ school interactions affect child ↔ school interactions). Influences from the increasingly distant layers influence the child's experiences indirectly. The inclusion of time indicates that both the child and the surrounding contexts are constantly changing, and thus that learning is always a dynamic process.

SOURCE: Adapted from Liben, L.S. (June 2014). *An Ecological Framework for STEM Learning*. Presentation at the National Summit on Successful Out-of-School STEM Learning, June, National Academy of Sciences, Washington DC.

A systemic approach to education policy that aligns with the ecosystem perspective considers the range of learning opportunities across settings and times.[8] Such an approach would ensure that learners have access to learning experiences that reflect and respond to young people's interests and prior experiences and connect to additional opportunities.

Examples of Connected STEM Learning Opportunities

Educational leaders in some communities are making concerted efforts to identify, diversify, connect, and broker young people's STEM learning opportunities across the learning ecosystem.

In the BRIDGE project (1996-2000),* New Mexico State University researchers worked with teachers across the school district to document how young people's home and community activities incorporated mathematical skills and knowledge. Educators used this documentation to design school programs that used young people's home skills and resources as starting points for academic work.

The Urban Advantage Program,** launched in 2004, led by the American Museum of Natural History in New York, is a collaboration between school districts and cultural institutions that encourages programmatic connections among family events, research using collections from museums and other informal settings, and the district-mandated 8th-grade exit project.

The HIVE project in Chicago*** was launched in 2012 to identify and connect the broad range of out-of-school programs available for youths to help families and youths locate interesting programs and to help programs broker ongoing opportunities for youths.

A particular strength of such coordinated efforts is to engage a more inclusive range of children in STEM, and to sustain their interest, participation, and learning over time.

---

*See http://math.arizona.edu/~bridge/ [May 2015].
**See http://www.urbanadvantagenyc.org/ [May 2015].
***See http://hivechicago.org/ [May 2015].

The Importance of Out-of-School Programs for STEM Learning

Following *Successful K-12 STEM Education*,[9] we identify three long-term, interrelated goals of STEM education: (1) increasing advanced training and careers in STEM fields; (2) expanding the STEM-capable workforce who serve as STEM educators, science communicators, medical assistants, computer technicians, and other STEM-related careers; and (3) increasing scientific literacy among all young people, supporting life-long interest and engagement with STEM. These long-term goals consist of many intermediate- and short-term goals, including learners' participating in STEM practices, developing learners' positive dispositions toward STEM, and creating social settings that promote life-long STEM learning. It is important to stress that STEM literacy is defined as involving far more than conceptual knowledge and skills: it also involves interest, reasoning, and understanding of real-world relevance.[10] These aspects of STEM literacy are not secondary goals: they are intrinsic and intertwined with understanding and engaging with STEM.

Although the majority of reform efforts that address the three broad goals of STEM education have focused on schools, children of school age spend only 20 percent of their waking hours in schools; the other 80 percent is spent outside of school, including in supervised out-of-school programs that meet after school hours, on weekends, and during the summer.[11] Strategies that support STEM learning, such as hands-on learning experiences, inquiry-based pedagogy, and connecting STEM to everyday life are widely applied in many out-of-school STEM programs.[12] Furthermore, out-of-school STEM programs leverage common structural features of out-of-school settings (e.g., hands-on activities, ungraded or unassessed activities, multiage groupings, fluid uses of time) to spark, sustain, and extend young people's interest, developing understanding, and commitment to STEM.[13] These findings suggest that STEM in out-of-school programs can be an important lever for implementing comprehensive and lasting improvements in STEM education.

The committee's review of current research and practice confirms that the evidence about learning in out-of-school programs, while promising, is not yet robust or consistent. This is not surprising for several reasons. First, many out-of-school experiences are short term and their effects will occur over time and across settings. Consequently, it is difficult or impossible to collect the downstream evidence of the program's impact. For example, the interest and skill developed in a science and engineering summer camp may later manifest itself in increased interest and achievement in an autumn science class, or at a different program the following summer. Second, because designers of out-of-school programs seek to engage, inspire, and broaden learning for young people partly by differentiating the programs from schooling, most avoid implementing tests and other familiar short-term ways of monitoring young people's learning. Third, the existing data on out-of-school programs frequently focus at the program level rather than the individual level. The program measures tend to be as diverse, local, and nonstandardized as the programs themselves. This specificity allows local programs to understand the programs' effect, but it simultaneously makes it difficult to aggregate the evidence across programs.

Historical Perspective on Evidence for Out-of-School STEM Learning

Out-of-school learning has a long history, dating back to the 18th century when institutions such as libraries, churches, and museums were seen as the main institutions concerned with public education.* However as an organized field, the out-of-school community is quite young. Recent years have seen an increase in research on how, when, where, and why children and youths learn across their days and over their lives. Although much has been learned, it is fair to say that much remains unknown.**

Our understanding of out-of-school STEM learning primarily comes from two forms of published knowledge—(1) studies that have been published in peer-reviewed academic journals, and (2) studies that are the result of internal or external evaluations of specific exhibitions or activities or other commissioned reports. Many research traditions and perspectives have contributed to what is known about out-of-school learning, including youth development, learning sciences, cognitive development, and informal learning.

Since 1980, research on informal STEM learning has increased dramatically. Investigations of STEM learning and engagement in out-of-school contexts have been published in many journals. There exists a substantial body of empirical work and scholarship that addresses the field of learning STEM in out-of-school contexts. However, there are notable gaps in the literature.

---

*Conn, S. (1998). *Museums and American Intellectual Life: 1876-1926.* Chicago, IL: University of Chicago Press.

**Peter, N. (2002). *Outcomes and Research in Out-of-School Time Program Design.* Philadelphia, PA: Best Practices Institute.

Although evidence of the effect of out-of-school programs is limited, a number of studies illustrate that out-of-school programs can contribute to young people's understanding of and interest in STEM. *Learning Science in Informal Environments: People, Places, and Pursuits*[14] details the many ways that learning STEM in out-of-school settings contributes to people's engagement with and pursuit of science learning. For example, out-of-school programs are well positioned to broaden participation in STEM learning by providing inquiry-based STEM experiences not commonly available in underresourced schools typically located in low-income communities.[15]

Out-of-school programs are likely to be taught by adults in the local community, thus providing important role models and community connections that can encourage pursuit of STEM learning.[16] In addition, the absence of high-stakes testing in out-of-school programs can allow for more flexibility and therefore inclusive approaches to STEM learning, which may encourage young people who do not yet see themselves as STEM learners.[17] Consistent participation in out-of-school programs has also been linked to performance in school and career choice. For example, studies have found that consistent participation in out-of-school programs leads to a narrowing of the achievement gap between young people from low-income and high-income families, better attendance, and more enthusiastic participation in school.[18] Retrospective and longitudinal studies of practicing scientists find that their experiences at home, in their community, or in other settings were at least as important as school for fueling their passion for and understanding of science.[19]

Out-of-school programs that contribute to the long-term, intermediate, and short-term goals of STEM education have three design features in common: they are *engaging*, *responsive*, and *make connections* across learning experiences. Engaging STEM learning experiences are an essential starting point. They attract children and their families by offering distinctive, well-designed activities that include interaction with STEM phenomena through visual media, the outdoors, hands-on explorations, exhibits, and other formats. Responsive programming taps into interests and understanding generated through prior experiences to optimize the relevance and accessibility of the program activities. Engaging and responsive programs can support inclusion, but their aggregate effects on young people require that they make connections across many learning opportunities. If a child deeply engages with engineering activities at home or in after-school programs, for example, but has no subsequent opportunity to build on those experiences in school or elsewhere, the longer-term value may be lost.[20] These three design features of productive out-of-school STEM programs are the basis for the criteria for identifying productive programs, which is the subject of the second chapter of this report.

Meeting the Demand for Out-of-School STEM Programs

Achieving the three goals identified above is not simple, but there are many existing programs, settings, and opportunities that comprehensive improvement efforts could leverage. For example, the number of young people enrolled in after-school and summer programs has skyrocketed over the past decade: currently one in five children participate in such programs.[21] Parents report that about 70 percent of the out-of-school programs available include STEM activities, and more than 50 percent of programs engage young people in STEM activities at least twice a week.[22]

Many organizations have been expanding STEM learning opportunities in out-of-school programs in recent years. For example, an increasing number of youth development organizations, such as 4-H, the Boy Scouts and Girls Scouts, and Boys and Girls Clubs embrace STEM as an important strategy for supporting youth in their intellectual, social, and emotional development. Expanded STEM learning opportunities can also be seen in the growth of citizen science programs and Makerspaces,[d] as well as an increased focus on STEM learning in public institutions, such as science centers, museums, and libraries.[e] Programs that focus on academic achievement and enrichment such as 21st Century Community Learning Centers also have begun to include STEM learning, with some 15 states currently having made STEM a priority focus.[23] There has also been an increase in the number of environmental science, math, and engineering camps; habitat restoration projects; after-school hobbyist clubs on such topics as robotics and astronomy; and multiday expeditions—such as fossil-hunting trips—that provide STEM learning opportunities.

In addition to program-level expansion, there are now more than 40 statewide after-school networks that support coordinated approaches to after-school programs, including 17 with a specific focus on STEM.[f] Various governmental, private, and corporate funders have undertaken a range of efforts to build the capacity of youth organizations to provide more robust and inclusive STEM learning opportunities. They have done so by building a broad and overlapping infrastructure of elements to support coordinated and high-quality settings and programs and by encouraging greater coordination of learning opportunities among schools and across out-of-school settings.

Despite the increase in programs, only one-third of the national need for out-of-school programing is being met by existing programs.[24] In addition, research has raised questions about the quality of STEM learning experiences in existing programs. A recent study of out-of-school programs in California found that most programs include STEM activities, but only a small proportion provide

[d]For more information about the program, see http://www.makerspace.com/ [May 2015].

[e]For more information, see the Institute for Museum and Library Service's STEM efforts at http://www.imls.gov/about/stem.aspx [May 2015].

[f]For more information, see http://www.statewideafterschoolnetworks.net/ [May 2015].

Developing an Infrastructure

Educators, funders, and governmental agencies have undertaken several notable efforts to create sustainable STEM learning infrastructure supports over the past two decades. Although none of these fully meet the national need, they illustrate a promising trend and foundation on which more comprehensive efforts can be built. The examples below illustrate those efforts.

Creating Statewide Coalitions for STEM Learning Opportunities Supported by the Charles Stewart Mott Foundation, over the past 12 years state representatives have been meeting annually to share strategies and undertake collective actions. These statewide after-school networks recently began to plan and build STEM-focused systems to provide more high-quality STEM learning opportunities that excite, engage, and inspire young people in their states. With additional support from the Noyce Foundation, the Mott state coalitions have worked to map STEM assets for after-school programs in their states, to leverage public and private funding, and to build good policies and practices to further after-school, summer, and expanded learning opportunities. For more information, see http://www.statewideafterschoolnetworks.net/ [May 2015].

Increasing Collaborations Between After-school Providers and Science Centers The Afterschool Alliance and the Association of Science-Technology Centers (ASTC) are working together to bring more high-quality STEM programs for young people to after-school programs. The initiative, announced in 2013 as a commitment to the Clinton Global Initiative, provides a series of conferences and meetings to create an "ASTC Community of Practice" that includes educators from science centers, museums, zoos, and planetariums and the providers of after-school programs to find ways to connect more ASTC members and after-school programs at the local level and to increase the quantity and quality of STEM in after-school programs nationwide. For more information, see http://www.astc.org/professional-development/communities-of-practice/ [May 2015].

Expanding the Reach of STEM Through Youth Organizations The Noyce Foundation in 2006 initiated a strategy to increase access to high-quality STEM learning opportunities through large national organizations whose leaders were already interested in providing science programs for children and youths but did not yet offer such programs on a large scale. For example, a series of grants to the National 4-H Council enabled 4-H to include substantial hands-on STEM programming for more than 1 million children and youths each year. Other grants to organizations, such as Girl Scouts, Girls Inc., and YMCA of the USA, have made it possible to reach millions more children and youths with high-quality STEM programs outside of school. For more information, see http://www.4-h.org/youth-development-programs/4-h-science-programs/ [May 2015], http://www.girlscouts.org/program/basics/science/ [May 2015], http://www.girlsinc.org/resources/programs/girls-inc-operation-smart.html [May 2015], and http://www.ymcanyc.org/association/pages/stem-science.-technology.-engineering.-mathematics [May 2015].

Creating Measures of Out-of-School STEM Learning There are several initiatives under way to develop measures of learning in out-of-school programs. For example, with initial support from the Gordon and Betty Moore Foundation, a team at the Lawrence Hall of Science, SRI International, and the University of Pittsburgh have been working to identify the factors that distinguish children who lose interest in science when they get to middle school from those who go on to become active science learners in high school and beyond. In a related activity, the Science Learning Activation Lab has developed measures of interest, curiosity, motivation, reasoning, and persistence in science, as well as appreciation of the value of science, responsibility for learning, and identity as a science learner. Other efforts to develop measures are discussed in the third chapter of this report. For more information, see http://www.activationlab.org/ [May 2015].

Aligning Support for STEM Learning Opportunities The STEM Funders Network is composed of more than a dozen private foundations that support STEM in both schools and informal settings. Facilitated by the Teaching Institute for Excellence in STEM, representatives from each of the foundations meet periodically to share ideas and develop collaborative strategies so that together they can have a deeper and longer-lasting impact than any one foundation might have alone. For more information, see http://www.tiesteach.org/solutions/stem-network-design/ [May 2015].

Supporting the Field of Out-of-School STEM Learning Supported by the National Science Foundation, the Center for Advancement of Informal Science Education (CAISE) provides an infrastructure for the out-of-school STEM education field. CAISE provides resources for practitioners, researchers, evaluators, and STEM-based professionals. It also facilitates conversation, connection and collaboration, and hosts searchable repositories of programs, evaluation reports, peer-reviewed research, and unpublished reports. For more information, see http://informalscience.org/ [May 2015].

Building the Capacity of Science Centers The DeWitt-Wallace Reader's Digest Fund, in collaboration with the Association of Science-Technology Centers, created Youth Alive! (Youth Achievement through Learning, Involvement, Volunteering, and Employment) to recruit teenagers from underserved local communities to work at science centers after school and during the summer. During its life from 1991 to 2001, 72 science centers received grants to start such programs, which soon became integral to the science centers' missions. Ten years after funding ended, the number of science centers with such programs has grown to 163, with many positive effects for both science centers and teenagers.* Many science centers reported increased cultural sensitivity among staff and increased integration of the institution with the local community. Increased school attendance, academic aspirations, and interest in STEM careers were found among participating teenagers. Efforts to provide professional development, staff training, and community partnerships continue through the YouthAlive! Regional Networks, which were created in 2000. For information about the regional networks, see http://www.astc.org/professional-development/youth-program-networks/ [May 2015].

*Sneider, C.I., and Burke, M. (2011). *The Legacy of YouthALIVE!* Washington, DC: Center for the Advancement of Informal Science Education. Available: http://informalscience.org/images/research/SneiderandBurke_LegacyofYouthAlive.pdf [February 2015].

opportunities for youths to participate in inquiry-based STEM learning.[25] Thus, there is a need to expand access to productive out-of-school STEM learning programs by improving existing programs and creating new ones.

Access to out-of-school STEM programs also remains a concern because STEM-rich out-of-school experiences are not evenly distributed.[26] Many children have their out-of school time carefully orchestrated by parents and families, who enroll them in programs and lessons, take them to museums and parks, and involve them in hands-on activities at home. And many of these activities involve STEM learning, either directly or indirectly. Yet not all parents and families have the time, the resources, or the information needed to access community resources to strategically organize their children's learning in activities outside of school. Policy makers can help address this basic inequity through policies that enrich, support, and expand high-quality STEM out-of-school learning programs.

CRITERIA FOR IDENTIFYING PRODUCTIVE STEM PROGRAMS IN OUT-OF-SCHOOL SETTINGS

Providing access to productive out-of-school STEM learning opportunities is key to enriching STEM learning for youths and children. As described in the first chapter of this report, productive programs are intellectually, socially, and emotionally engaging. They reflect and develop young people's interest in and understanding of STEM and provide connections to the broader ecosystem of STEM learning and career pathways. In detailing what counts as productive in this chapter, we also pay particular attention to how programs can actively seek to broaden participation of youths from communities historically underrepresented in STEM fields.

The criteria for identifying productive out-of-school STEM programs are derived from syntheses of research and practice in the fields of youth development,[27] learning science in informal environments,[28] and connected or cross-setting learning.[29] In discussing the supporting evidence for the criteria, we provide examples of how they can operate in practice. Our criteria fall into three categories:

1. **Productive programs engage young people intellectually, socially, and emotionally.**
- They provide first-hand experiences with phenomena and materials.
- They engage young people in sustained STEM practices.
- They establish a supportive learning community.

2. **Productive programs respond to young people's interests, experiences, and cultural practices.**
- They position STEM as socially meaningful and culturally relevant.
- They support collaboration, leadership, and ownership of STEM learning.
- They position staff as co-investigators and learners alongside young people.

3. **Productive programs connect STEM learning in out-of-school, school, home, and other settings.**
- They connect learning experiences across settings.
- They leverage community resources and partnerships.
- They actively broker additional STEM learning opportunities.

Our review of the research suggests that productive out-of-school STEM programs demonstrate a dynamic and interwoven relationship among these three sets of criteria.[30] For example, productive STEM out-of-school programs that intentionally leverage young people's interests help make explicit the connections between STEM experiences across multiple settings[31] and help them see the relevance of those experiences to their daily lives and future careers, which can deepen their intellectual, social, and emotional engagement with STEM.[32]

Engage Young People Intellectually, Socially, and Emotionally

Research suggests that intellectually engaging STEM programs provide young people with first-hand, materials-rich, and place-based learning opportunities that involve processes of scientific or engineering investigation and practice.[33] These opportunities help to make STEM a living field of activity and allow for multiple modes of learning (including visual and tactile).[34] Out-of-school STEM programs can provide young people with the time, community, and support needed to engage in STEM practices for a sustained period. For example, they can provide young people the opportunity to develop and pursue STEM questions or ideas that have personal meaning over time (whether hours, days, or weeks) in ways that can encompass the full range of STEM practices—from problem to solution or from question to explanation.[35]

Because learning involves intellectual, social, and emotional engagement, it is best supported in social environments that inspire young people to participate, that offer opportunities to contribute to a shared endeavor, and that provide the necessary social supports that allow young people to stretch themselves intellectually, socially and emotionally.[36] The key attributes of supportive out-of-school programs that lead to young people's meaningful participation and development include[37]

- physical and psychological safety;
- opportunities for belonging;
- support for efficacy and mattering (meaningful involvement);
- appropriate structure;
- opportunities for skill building;
- integration of family, school, and community efforts;
- supportive relationships; and
- positive social norms.

First-hand Experiences with STEM Phenomena and Materials

Research demonstrates the power of learning through first-hand experience with phenomena and materials.[38] In STEM learning, first-hand experience is often equated with "hands-on." But first-hand is more than hands-on: it can include place-based investigations, computer-based studies of complex systems, projects that explicate the relationship between STEM and society, as well as hands-on explorations of physical properties and materials. First-hand means providing young people direct engagement with questions, contexts, and data in all of its relevant forms.

The productive out-of-school STEM programs we reviewed provided young people with such first-hand experiences. They included opportunities to care for small animals in a local community zoo; data collection activities involving mapping of neighborhood trees and interviews

with community residents; tabletop investigations of light and color; and design and engineering activities to fabricate digital clothing. They also included Web-based research and data collection and visits to local community STEM-rich settings.

One example of a program that provides first-hand experiences with materials and phenomena is the California Tinkering Afterschool Network, a collaboration among five STEM-rich organizations working in partnership with after-school programs in urban and rural California communities.[39] At one partner site, the Community Science Workshop (CSW) in Watsonville, young people drop by to use the workshop tools and materials to build objects of their own choosing motivated by needs from home (such as a fountain for a garden), school assignments (such as a working trebuchet), or models found in the workshop (such as a wooden tortilla press or a Rube Goldberg machine). Young adult staff support young people's ideas and teach them how to use the tools, to plan and measure, and to troubleshoot their designs.

At CSW, young people learn about materials and phenomena by working directly with them. For example, a group of three girls who were regular drop-in participants signed up for a CSW summer field trip to a local lake, having decided that they wanted to build a canoe that they could use on the trip. First, they had to determine what kind of materials would both float and carry the weight of at least two people, choosing from the low-cost materials available in the workshop. Assisted by a facilitator, the girls investigated workshop materials. After initial exploration, they began to experiment with the use of duct tape as a material that might provide a lightweight but waterproof skin for the canoe. Over several days, they designed and tested different ways to layer the duct tape to attain the desired characteristics of abrasion, puncture, and tear resistance (e.g., weaving strips together as in a fabric, layering them on top of each other as in a roof). After creating and testing several small-scale prototypes, they decided to alter their plans a bit by creating two canoes that would support passenger weight on their frames, using the duct tape as a material to keep the water out but not to bear direct weight. The canoe frames were created using flexible PVC (polyvinyl chloride) pipe, connected by cross bars and a platform in the style of a catamaran, and finally wrapped using the woven tape. A few days later, the girls indeed successfully used the canoe on the field trip. This example illustrates what first-hand engagement with materials and phenomena can look like in out-of-school settings: it can be purposeful, iterative, and collaborative.

Engagement in Sustained STEM Practices

Research has demonstrated that one of the best ways to learn STEM is to engage in the practices of doing STEM.[40] Direct involvement in STEM practices gives young people an appreciation of the wide range of practices that are used to investigate, model, and explain natural phenomena and the man-made world.[41] STEM practices include asking questions and defining problems; developing and using models; planning and carrying out investigations; analyzing and interpreting data; using mathematics and computational thinking; constructing explanations and designing solutions; engaging in argument from evidence; and obtaining, evaluating, and communicating information.[42] Scientists and engineers fluidly and iteratively move back and forth among these practices, and they carry out activities that might involve multiple practices at once. Through direct engagement in scientific and engineering practices, young people can experience and understand STEM as a powerful approach for exploring, learning about, and interpreting natural phenomena and the constructed world; they also learn how STEM knowledge develops.

Techbridge, for example, is a nonprofit organization that provides after-school and summer programs that aim to inspire girls in underserved communities to discover a passion for STEM by engaging them in STEM practices through hands-on learning. Through collaborations with the Girls Scouts of the USA, YMCA, and others, Techbridge provides training to the adults who serve as role models and facilitators. The after-school and summer programs engage girls in real-world applications of science, engineering, and mathematics; support career exploration with role models; and promote leadership. Participants engage with STEM practices on a regular basis as they work on projects over several weeks that involve posing problems, designing solutions to problems as they arise, testing designs while conducting investigations, revising designs based on their findings, and communicating their findings.

Research on Techbridge programs documents how its activities provide a motivating context for girls to engage in STEM practices. For example, one Techbridge project involves girls hacking or repurposing an item they have selected from a local community-run thrift shop. The girls begin their project by choosing an item to hack and sketching out an illustration of what they plan to do (e.g., sketching out where they might add LEDs [light-emitting diodes], where they might add small speakers, what they might take apart and put together in new ways). One participant planned to add a sensor at the bottom of the Buddha-shaped coin bank, so that the mouth would light up whenever a coin was added. Over a period of weeks, as she worked on her project, she came to realize that the sensor pad at the bottom of the bank did not register every coin as it was dropped in the bank. With additional investigation of her materials (and assistance from a facilitator), she realized that she could perhaps get coins being fed through the bank coin-slot to complete a circuit with metal touching metal, and that this circuit, in turn, could make an LED light up. After successfully testing whether the coins contain enough of the right type of metal to complete a circuit, the girl created and attached the circuit to the bank's coin-slot, and it worked. This project illustrates what STEM practices (e.g., planning and carrying out investigation, developing and using models, and designing solutions) can look like in out-of-school settings: in this case, it was creative, whimsical, and personally designed.

Supportive Learning Communities

In productive STEM learning environments, young people are encouraged to develop their own questions, to devise ways of investigating and addressing those questions, and to share the results of their inquiries, which will often be tentative. This type of experience is a fundamental part of doing science and being scientific.[43] Young people who feel supported to explore the unknown are more likely to attempt explanatory modeling and to persist after experiencing a moment of failure, which can lead to a moment of new insight.[44] Research shows that socially supportive contexts are linked to such outcomes as increased pro-social behavior and school achievement.[45] Thus, thoughtfully designed supportive learning communities may be key to young people's STEM learning in out-of-school programs, and they may be particularly important for broadening participation in STEM for young people from historically underrepresented communities.

An example of such a supportive learning community and its role in positioning children in grades 3-5 for success comes from Communities Educating Tomorrow's Scientists (COMETS), an after-school program in West Virginia funded by the U.S. National Science Foundation. COMETS was part of a larger initiative that demonstrated sustained participation and interest in STEM among middle school students in comparison to their matched counterparts, for whom participation and interest declined.[46] In these programs, staff frequently made accommodations for students who had particular areas of expertise or interest; who were tired at the end of a long day, or perhaps struggling with family issues and in need of interpersonal care; or who needed to express themselves in their own unique ways.

AREAS FOR FUTURE RESEARCH: There is a need for more detailed accounts of how STEM learning in out-of-school settings emerges through the intertwining of intellectual, social, and affective dimensions of learning environments.

An example of social support in this context comes from an activity about hurricanes. After watching a short video about hurricanes as part of learning about meteorology, a program facilitator sat down with the children in chairs circling a round table and asked them to share with each other what they knew about hurricanes. "They have strong winds," said one child. "They can blow your house down," another said. The next child, a boy about 8 years old, stood up and began to silently but energetically spin around the table like a hurricane. Eventually his "orbit" brought him back to his starting point and he collapsed into his chair. "I'm going to build my house on stilts," he said, so that his house couldn't be flooded. He switched to talking about how dogs fared in hurricanes. The facilitator smiled encouragingly at the child while he spoke, responding to his comments with "Really?" and "Uh-huhs." He let him finish his thought on dogs and then quietly suggested that they ask the next child what she knew about hurricanes. All eyes turned to the next child at the table. This combination of sensitivity to children's moods and accommodation of their interests in the context of STEM activities corresponds to goals to motivate children and promote their interest in STEM.[47]

Respond to Young People's Interests, Experiences, and Cultural Practices

Many young people experience STEM as an abstraction that appears to have little connection with their daily lives.[48] Commonly, young people's ideas about STEM reflect cultural models that include images of obsessive genius scientists working lonely late night hours in their laboratories.[49] Young people are less likely to understand STEM as a collaborative and team-based activity, they seldom picture STEM practices as involving artistic and detailed representations of the natural world, and they consistently associate being good at STEM with natural ability rather than hard work.[50] Such cultural models make STEM less appealing to many young people who envision their future life's work as addressing significant issues in their communities. A major goal of STEM education therefore is to help young people to understand the relevance of STEM to the worlds they know, so they can understand the utility and value of STEM and how it is situated in meaningful social contexts.[51]

There is a relationship among prior experiences, beliefs, relevance, and engagement in education.[52] When young people recognize a question, problem, or strategy as meaningful, they are more likely to become interested in it.[53] When they are interested in the idea or topic, they are more likely to pursue it.[54] When they believe that a skill will be of value to them in their immediate context, however they define it, they are more likely to persist in learning it.[55] Young people who are supported to persist and succeed and to reflect on their tenacity, are more likely to apply themselves and, indeed, to succeed.[56] Understanding how to make out-of-school STEM responsive to young people's prior interests and experiences so that they can see STEM as meaningful and relevant to their own experiences and aspirations may be especially important for youths from communities historically underrepresented in STEM fields.[57] Girls and youths from economically marginalized communities, including immigrant communities, are frequently treated, explicitly and implicitly, as less capable in STEM and therefore may approach STEM with hesitation or even antipathy.[58] Ensuring access to high-quality, personally relevant, and responsive out-of-school STEM programming may be a valuable strategy for addressing equity issues in STEM education.

STEM as Socially Meaningful and Culturally Relevant

Recent research on the relationship between supportive and culturally responsive out-of-school STEM programs and STEM learning, and more detailed accounts of what culturally responsive STEM out-of-school learning looks like and leads to are needed. Yet there are compelling accounts demonstrating that when programs explicitly connect STEM to recognizable problems in a community and leverage the participants' cultural resources and practices, the possibilities for STEM learning experiences are expanded.[59] Such cultural practices include discourse patterns (e.g., overlapping speech patterns, hybrid bilingualism, uses of metaphors, and ways of questioning) that can be engaged to support scientific argumentation,[60] familiar home skills and practices that can be engaged (e.g., sewing, banking, carpentry, or fixing things) to encourage young people's participation and skill sharing,[61] and belief systems that can be engaged to support observations and analysis of natural phenomena.[62]

Supporting young people's appreciation of how STEM is relevant to important questions and problems can engage youths who may not self-identify as STEM learners but are committed to social or community issues. Situating STEM learning in relevant settings and contexts can also assist young people who may feel cultural dissonance between current cultural meanings of science, for example, and their personal systems of belief (e.g., religious) or family histories (whether any family role models have ever worked in the sciences). Out-of-school STEM programs that situate STEM in relevant settings and contexts treat young people as knowledgeable and capable, thus supporting them intellectually, socially, and emotionally to fully participate, contribute, and develop as members of the STEM learning community.[63]

Native Science Field Centers, a program developed by Hopa Mountain and Blackfeet Community College, serves as an example of how STEM learning experiences can be designed to be socially meaningful and culturally relevant. These centers strive to create relevant environments in their year-round programs for young people. Their programs explicitly connect traditional culture and language with Western science: for example, young people engage in environmental observations in their own communities, learning empirical observation and recording techniques as well as tribal traditions related to the natural environment.

The program depends on community involvement. An advisory board ensures that program developers are implementing traditional knowledge in an appropriate way and provides guidance and support in developing curriculum materials and finding resources. Parents, teachers, and tribal elders contribute by donating materials for the projects, sharing their knowledge, and volunteering their time. This collective effort leads to activities that bring together cultural traditions and STEM practices. For example, a harvesting activity at the centers begins with participants huddling in a circle, reciting a prayer in their language, and making an offering of tobacco—traditions meant to make the youths aware of the reciprocal relationship they have with mother Earth. Research indicates that engaging in such activities helped participating youths build self-efficacy in STEM and confirmed for them the value of the cultural knowledge of their communities.[64] This example illustrates how socially meaningful STEM experiences can engage young people in STEM practices and learning.

Supporting Collaboration, Leadership, and Ownership of STEM Learning

Research shows deep links between identity development and learning,[65] illustrates the importance of engaging youths as both leaders and learners,[66] and demonstrates the significance of addressing young people's agency in their learning.[67] Participation in collaborative communities of practice[g] can be critical to the emergence of identities, and, specifically, to the development of practice-linked (or domain-specific) identities[68] and a sense of belonging.[69] Researchers have documented the roles of STEM communities of practice in shaping commitments to STEM learning.[70]

Collaborative learning strategies, including problem-based learning approaches, may provide especially flexible contexts for allowing young people to leverage their own strengths, interests, skills, and even networks to ensure team success. For example, if one person's data skills and another person's facility with engaging older adults help a team successfully interview community residents to investigate and later communicate health conditions in an urban neighborhood, a program may have positioned both young people to develop productive STEM learning identities.[71] Project-based learning may be an especially productive strategy for learners to develop and evolve in their roles in communities of practice, providing young people the opportunity, over time, to take on new roles as the project progresses.[72] Because of the time-dependent and often site-specific nature of project-based learning, it may be well suited to out-of-school settings that can allow for extended investigations.

[g]A community of practice is a group of people who learn from each other and has an opportunity to develop themselves personally and professionally by sharing information and experiences with each other.

For example, the Green Energy Technology in the City (GET City) Collaborative provides a series of after-school and summer engineering design experiences for youths in the Lansing, Michigan area.[b] GET City programs provide participants the opportunity to explore energy issues; engineer creative solutions to energy problems; and educate peers, community members, and local organizations about energy issues.

One GET City project involved investigations of urban heat islands and their effects on community health.[73] The multifaceted nature of the project, which incorporated research, engagement with community members, creating meaning, and presenting results at a town hall meeting, created opportunities for youths who did not self-identify as STEM learners to find meaningful ways to engage in STEM. Although they may have become engaged in the GET City activities through an identity as a community advocate—with an interest in interviewing community members and presenting at the town hall—participants worked with STEM concepts, data, analysis, and data representations, in the process coming to see themselves as capable in STEM.

Position Staff Members as Co-Investigators and Learners with Youths

To create productive STEM out-of-school programs that reflect the criteria described above, skilled and caring adult support is essential.[74] Supportive relationships involve adults who come to know and to recognize the strengths and interests of program participants and empower them to identify and pursue their own meaningful questions.[75] These relationships can develop when staff members work alongside young people as co-investigators, asking "what-if" questions and recasting "failure" as a fundamental part of learning and scientific endeavors. Supporting youths to take ownership of their learning may be especially important in out-of-school settings, where young people are developing new interests and deepening existing ones that can be further pursued in other settings including school.

[b]For details, see http://getcity.org/ [May 2015].

One example is the Oakland-based Youth Radio, a program that allows young people (ages 14 to 24) to take on roles as reporters of science stories and developers of technology.[i] The program is designed around fluid roles and relationships among participants and between young people and adults. Program activities include classes and workshops, peer mentoring, and paid reporting activities done in collaboration with adults to produce reports for National Public Radio. The program uses what it calls "collegial pedagogy"—a relationship of shared responsibility—to support young people's advancement from novice to expert as they develop their journalistic capacities as well as understanding of the subject matter (often science, engineering, health, or technology).[76] The Youth Radio program builds a collegial pedagogy by creating a context in which adult experts and young people are mutually dependent on each other's skills and perspectives. The program creates such an environment through joint framing of an issue, youth-led inquiry, and distributed accountability.

One Youth Radio program relevant to STEM is Young Radio Investigates (YRI). YRI engaged youths in collaborations with scientists and radio producers to examine data on a personally relevant STEM issue and report the results in a major news outlet. With guidance from scientists, the participants collected and analyzed primary and secondary data. In addition, the participants worked closely with producers to identify credible sources and translate findings for media outlets. One participant who volunteered to develop a story on a sensitive public health issue worked closely with the producer to determine what aspects of the health issue should be the focus of the story and what question the story should address. After a series of discussions, the participant suggested that the story be framed around the neurological aspects of the PTSD (post-traumatic stress disorder), because she has struggled with similar neurological issues. The producer agreed it would be an interesting angle for the story and helped develop the idea into a full news story. An evaluation of the program found that participants learned STEM concepts, developed more positive attitudes toward STEM, and acquired technical skills related to computer programing.[77]

Connect STEM Learning in Out-of-School, School, Home, and Other Settings

Researchers have begun to develop strategies for understanding and documenting how learning develops, fluctuates, and deepens across settings and over time.[78] A growing number of studies demonstrate how young people bring STEM understanding and practices developed in one setting to another, including between home and school,[79] between school and out-of-school activities,[80] between home and out-of-school activities,[81] and across out-of-school settings.[82]

Productive out-of-school STEM programs can help young people understand how their out-of-school experiences build on, connect with, and support continued learning and activity in other

[i]For more information, see https://youthradio.org [May 2015].

settings, including school.[83] Making these connections provides context and meaning to young people; failing to make these connections can have negative consequences for their interest and growing expertise in STEM. Collaborations or institutional partnerships among organizations have the potential to facilitate explicit connections between school and out-of-school programs,[84] monitor youth development across a wide array of settings,[85] and build networks of opportunities that are brokered to advance young people's engagement with STEM.[86]

AREAS FOR FUTURE RESEARCH: Research on the design of out-of-school programs is needed to better specify how culturally responsive and relevant out-of-school STEM learning experiences affect the short-term and long-term learning trajectories of young people, especially young people from underserved groups.

Although all young people would likely benefit from more brokering of learning opportunities, young people from economically marginalized communities, rural communities, or immigrant communities whose parents may not have access to or awareness of possible pathways and opportunities, may need more active brokering.[87]

Connecting Learning Experiences Across Settings

Historically, designers of STEM out-of-school programs have struggled with how to be or not to be "school-like." The unproductive dichotomy of school or not school has led to dilemmas about how active a role adults should play in supporting young people's learning, how sequential and coherent program activities should be over time, and when and how to introduce academic and disciplinary language and terms.[88] Too often the result is a belief that out-of-school learning should get adults "out of the way," prioritize individual moments of engagement over a coherent sequence of experiences, and keep academic or advanced language "out of the picture." When taken to the extreme, these approaches can shortchange possibilities for student learning and development. For example, it is well established that individuals learn best when supported by caring others. The goal is to position adults as active and responsive supports of student-directed learning. The dichotomy between school and out-of-school also negatively affects public perceptions of the significance and value of out-of-school settings and programs.[89]

Evolving understanding of learning beyond the classroom and of the importance of academic success to the well-being of youths and their paths to adulthood have challenged this dichotomy.[90] It is clear that young people benefit by becoming aware of how particular skills or understandings in one setting, such as an after-school program that has young people investigating local waterways, are relevant in another, such as classroom engagement with scientific practices.[91] Productive STEM out-of-school programs are not stand-alone or destination points but rather are points on a journey, recharging stations where young people can replenish, expand, and deepen their engagement with STEM learning.[92]

An example of a program that aims to provide ongoing connections among STEM experiences is the University of Pennsylvania's Penn Academy for Reproductive Sciences (PARS).[j] PARS is a 6-week Saturday program that offers girls in grades 10-12 the opportunity to explore their interests in reproductive health and research science and to learn directly from top professionals in these fields. PARS participants deepen their understanding of comparative developmental biology through hands-on laboratory experiences. Participants are asked to analyze scientific literature and discuss ethical scenarios related to their research experience.

The weekly lessons were designed to reinforce and create connections with disciplinary concepts taught in the participants' high school biology classes by taking abstract concepts (such as the structure and function of DNA, inheritance and variation of traits, and genetic diseases) and providing opportunities to use this knowledge as a foundation for experiments in a laboratory and in patient care in a clinical setting. The PARS program has built a network of teachers and youths who tell program staff about their classroom curriculum in science and biology and individual life experiences that might be relevant. Opportunities for participants to pursue their interest and extend their learning are also made available to alumnae through summer research and clinical internships. Thus, PARS makes connections through direct communication with teachers and provides tangible resources. This example illustrates how out-of-school programs can make connections explicit without subordinating the out-of-school learning experience to the school agenda.

Leveraging Community Resources and Partnerships

As ecosystem perspectives of learning continue to gain traction, more community organizations are seeking to build partnerships that can support and even track young people's STEM learning across settings.[93] This systemic view of learning includes multiple parties in a community who collectively seek to expand opportunities for STEM learning.

A number of studies have investigated the relationship between partnership structures and outcomes[94] and provide useful insights on how best to choose partners, establish clear lines of work and communication, and avoid pitfalls. Few studies, however, have investigated outcomes for youths. An exception is a recent dissertation study of museum-school partnerships,[95] which found that outcomes for participants were optimized when teachers and museum professionals collaboratively designed coursework that incorporated the instructional practices and instruments of both learning environments. In particular, when the coursework was organized in such a way that the young people were asked authentic STEM questions, were given authentic tasks to do, and their answers and products were taken seriously by their teachers and the museum professionals, the

---

[j]PARS is modeled after Dr. Theresa Woodruff's Oncofertility Saturday Academy, which focuses on exposing a diverse group of high school young women to the basic sciences. For more information, see http://irm.med.upenn.edu/science-impacting-the-clinic/education-outreach/pars/ [May 2015] and http://oncofertility.northwestern.edu/oncofertility-saturday-academies [May 2015].

young people were strongly motivated to do the work. This study suggests the potential of productively organized partnerships for creating new types of learning opportunities, not available without the partnerships, which can motivate and inspire youths to engage in STEM learning. The findings of the study also play out in practices in a number of communities including in New York City where the Urban Advantage program[k] has created partnerships among the public school system and the city's science institutions (e.g., museums, zoos, and science centers) in order to accomplish a number of goals, including connecting STEM learning in school and in out-of-school learning settings and providing professional development for educators.

One project that aims to leverage productive partnerships is SYNERGIES in Portland, Oregon, whose partners include the Parkrose School District; Oregon State University; 4-H Youth Development (Portland Metro Group); Math, Science, Engineering and Achievement Program (MESA); Girls, Inc.; Oregon Museum of Science and Industry; Multnomah County Library; Pixel Arts; and the port of Portland.[l] The project is designed to create a better, more effective, community-wide STEM education system for low- to moderate-income early adolescents living in Parkrose, which is a diverse, underresourced neighborhood in East Portland, Oregon. Together, the partners are developing a comprehensive, community-wide plan to improve youth STEM learning in Parkrose, both in and outside of school. SYNERGIES' staff and its partners work to ensure that each of the STEM learning opportunities in Parkrose interconnect, and that every STEM education provider knows what other Parkrose educational providers are doing, as well as what the youth in their programs are doing and what interests them.[96]

An example of the connected learning experiences facilitated by SYNERGIES is the partnership between the Parkrose Middle School science program, the Portland Port Authority (which includes the airport), and a major after-school program (Schools Uniting Neighborhoods program). Educators from all three groups, facilitated by the SYNERGIES community coordinator, developed integrated experiences focused on engaging youths in engineering practices. Connections across stakeholders such as these have led to STEM learning offerings that are more interrelated and synergistic.[97] In addition, a key asset that the SYNERGIES project has brought to the Parkrose community is the ongoing collection of data about youth interest and participation.

---

[k]For more information, see http://www.urbanadvantagenyc.org/ [May 2015].
[l]For more information, see http://education.oregonstate.edu/book/synergies-parkrose-community [February 2015].

Brokering Additional STEM Learning Opportunities

Brokering learning across settings is an important strategy for promoting greater diversity among STEM learners.[98] By brokering we mean actively identifying opportunities and networks that can assist youths in choosing activities. Just as a real estate broker selectively identifies potentially interesting properties for prospective homeowners, so can brokering help young people and their families become aware of potentially interesting choices and opportunities, and how to prepare for them.

In most communities there are STEM resources available to youths in or near their homes.[99] All young people need support in understanding how to navigate these possible learning experiences in order to advance and diversify their developing interests and skills.[100] Youths and families who have been historically underrepresented in STEM may especially benefit from more explicit brokering of these opportunities.[101] Active brokering can include directing young people to more advanced programs, helping them to secure internships or apprenticeships, and introducing them to professionals and other key individuals. It might also include creating opportunities for them to express their emerging STEM knowledge through leadership in clubs and other settings. Brokering can expand the personal networks of young people[102] and help them navigate educational requirements and expectations.[103] To be effective at brokering, out-of-school STEM program leaders need to be aware of opportunities for STEM learning in their communities. This may require them to develop relationships with other program leaders, including K-12 teachers, as well as with parents. Brokering can also be facilitated by community-level maps of STEM learning assets.

The Maine Mathematics and Science Alliance, in partnership with Maine 4-H and local organizations, has created STEM Guides, people who broker STEM opportunities for young people. The program focuses on creating links between rural youths ages 10-18 and the broad array of STEM resources that are available to them. The program recruits and trains a small number of local adults to become the STEM Guides, whose job it is to link youths to resources, particularly those resources that support individuals' developing interests. The STEM Guides also use proven community-based dialogue formats (such as Teen Science Café[m]) to connect local STEM-related professionals with young people who want to know about their fields.

For example, in a rural town of 4,000 people, a STEM Guide met with a 16-year-old youth and discovered he was interested in engineering. At the STEM Guide's suggestion, he joined the local youth leadership team to organize science cafes for local teens, and he helped select the first speaker, a design engineer. As a next step, the STEM Guide told him about an engineering summer camp for juniors, and when he was old enough he asked her for a letter of recommendation

[m]For more information, see http://teensciencecafe.org/cafes/teen-science-cafe-for-me/ [May 2015].

AREAS FOR FUTURE RESEARCH: Research is needed to better specify and understand the ways in which learning develops across formal and informal settings, leveraging community resources and partnerships.

to attend. He was accepted and thoroughly immersed himself in the experience. This youth is now planning to study engineering at the Rochester Institute of Technology. This example illustrates the ways in which adults can organize opportunities for youths to support ongoing engagement, learning, and commitments to STEM.

Cross-Cutting Issue: Staff Capacity

Whether or not a program embodies the criteria for identifying productive out-of-school STEM programs depends not only on the design of the programs, but also the actions of the frontline staff who work directly with participants. The preparation of frontline staff varies greatly. The frontline staff have backgrounds in a wide range of fields, including education, social work, sociology, urban studies, art, science, engineering, mathematics, and history. They may have high school diplomas, associate's degrees, bachelor's degrees, master's degrees, teaching certificates, social work licenses, or doctoral degrees.[104] There is also little consistency across programs in terms of job titles and responsibilities. Developing a high-quality out-of-school STEM workforce is complicated by the high turnover rate among frontline staff.[105] The diversity of staff backgrounds, education, and responsibilities, along with the high turnover rate makes it difficult to develop a high-quality workforce and to design effective professional development activities.

Effective professional development for out-of-school STEM facilitators and instructors needs to cover many areas: presenting ideas and concepts with a clear rationale for their importance, demonstrating new practices, taking advantage of staff experience and expertise, offering opportunities for practice and feedback, providing ongoing support and follow-up training, linking staff members with mentors, using planning time to cultivate collaboration among staff, and augmenting training time with resources and materials.[106] In addition, effective professional development provides educators with opportunities to learn about STEM disciplinary content and practices, as well as theories of child and youth development, in order to develop positive relationships with and empower youths, to decrease risk factors and maintain safe learning environments, and to implement age-appropriate activities. Professional development also prepares frontline staff to value cultural and ethnic diversity, to interact with families, schools, and communities, and to serve as professional role models, while integrating staff interests and input into all activities.[107]

4-H is one out-of-school STEM provider that has focused on improving the capacity of its staff members to facilitate productive learning experiences. The 4-H commitment to improving the STEM skills of America's youths has been present during the organization's 110-year history. Building on its history of hands-on science education, in 2007 4-H partnered with the Noyce Foundation to develop a nationally recognized youth development approach to STEM in out-of-school settings. A key aspect of this partnership was to create a professional development strategy to prepare state and local 4-H educators and volunteers.

4-H has developed a suite of materials[n] that can be used by state and local 4-H staff to train the 4-H science volunteers who facilitate education activities. The materials have been designed to be appropriate for training the volunteers who come from a wide range of educational and professional backgrounds, and they can be tailored to the needs of the local volunteers. Included in the materials are resources for building an understanding of quality STEM programs and for implementing professional development. The resources designed for building an understanding of program quality focus on what educators need to know about inquiry-based learning and further develop their understanding of the STEM concepts and positive youth development practices that frame 4-H STEM programming. The implementation resources are designed to provide strategies for 4-H staff to recruit, retain, and prepare volunteers.

[n]For example, see http://www.4-h.org/resource-library/professional-development-learning/science-training-guides-resources/ [May 2015].

EVALUATING OUTCOMES AND GENERATING NEW KNOWLEDGE

Evaluation is key to improving the overall quality of out-of-school STEM programs and to understanding how they contribute to the learning ecosystem.[o] Evaluations can inform program developers, researchers, policy makers, and the public as to what out-of-school STEM programs contribute to interest and learning. They can also provide information about the broader context of STEM learning in a community. In this chapter, we describe the complex nature of evaluating the outcomes of out-of-school programs, and what can be done to provide a clearer picture of what programs work best under what circumstances for whom, and how the programs fit into the larger STEM learning ecosystem. The chapter provides a framework to guide evaluation efforts.[p]

The Role of Evaluation

Evaluation has many purposes, including for continuous improvement, accountability, informing management, and demonstration of value. And it can take many forms, including one-time studies, ongoing cycles of data collection and reflection, and participatory evaluation. It can marshal the entire methodological toolkit available in social science and educational research, including multiple study design options and data collection methods. Evaluation efforts can include a range of study designs—with quantitative, qualitative, and mixed data collection methods—often done in collaboration with either in-house or external evaluation experts.

With all the possibilities for how evaluations can be used to document program implementation and outcomes, decisions about evaluation design and execution need to consider three elements: (1) the program's design (how the program is supposed to work, for whom, and with what resources), (2) the larger policy environment in which it is being operated, and (3) the most current knowledge in the field of evaluation itself. For example, deciding when to use different evaluation approaches is related to the maturity and focus of the program and the goals of the evaluation. Evaluations of new initiatives or programs may best be focused on the qualities of the program's

[o]Broadly speaking, evaluation of an out-of-school program is the systematic process of collecting information (data) to enhance understanding of how a program is operating and inform decisions.

[p]This chapter draws heavily on the papers by Barron and by Hammer and Radoff; see Appendix B.

design and implementation with an emphasis on formative feedback from participants' about the program content and pedagogy. Then, after a program is more stable, an evaluation could begin to focus more on whether the program is achieving its expected individual-level outcomes, which might be done with the structure of a more formal, summative evaluation. Once a program has undergone summative evaluation, it may be appropriate to conduct a comparative evaluation to understand a program's relative strengths and weaknesses in contrast to similarly designed programs or relative to programs that serve similar participant populations.[108]

The current climate of evidence-based policy and decision making increasingly requires that programs demonstrate their intended outcomes. In the field of education, broadly, funders, policy makers, and the public expect to see evidence of learning. Consequently, evaluations of education programs typically focus on individual learning assessments, where learning is defined in terms of gains in specific knowledge or skills.[q] How these outcomes are measured depends to a considerable degree on how a program's designers have defined learning outcomes and the factors affecting them. How do young people learn STEM? What does learning look like in action? What factors contribute to learning? The answers to such questions affect how evaluation studies define and measure learning.

[q]We note a useful distinction made by researchers between assessment and evaluation. "The educational research community generally makes a distinction between assessment—the set of approaches and techniques used to determine what individuals learn from a given instructional program—and evaluation—the set of approaches and techniques used to make judgments about a given instructional program, approach, or treatment, improve its effectiveness, and inform decisions about its development. Assessment targets what learners have or have not learned, whereas evaluation targets the quality of the intervention." (National Research Council, 2009, p. 54). Therefore, assessment of learning can be an element in the evaluation of a program, but it is not necessarily the only element that determines whether a program is productive.

Evaluating a STEM Learning Ecosystem

Evidence regarding which out-of-school programs support STEM learning and stimulate interest in STEM, how they do so, and for whom and under what circumstances has been slow to emerge due to the complex nature of STEM learning, the wide variation in the nature of out-of-school programs, and the quality of evaluations. Evaluations of out-of-school STEM programs are challenged by a number of theoretical and practical factors. We emphasize the accumulation of experiences, change at multiple levels, the idiosyncratic nature of learning, and additional evaluation challenges because they were the focus of discussions at the National Summit on Successful Out-of-School STEM Learning, they were highlighted in the background papers commissioned for this report, and they have been cited in major reports on out-of-school STEM learning. Although these same issues also create challenges for evaluating learning in all settings, we focus on what they mean for out-of-school programs.

Accumulation of Experiences

The success of out-of-school STEM programs depends on the possibilities they create for young people to expand, deepen, and reinforce their cumulative STEM experiences. Since a wide array of activities, people, programs, material resources, and facilitators sustain engagement, the accumulation of learning opportunities usually accounts for development of expertise and interest (though, occasionally, one powerful experience is transformative).[109] A single experience may not have an immediately recognizable or detectable effect on knowledge or interest, but it may have a relatively profound effect if it serves to orient, inspire, or motivate a young person to be open to new STEM learning opportunities.[110] It is very difficult to know whether an after-school hike, an intriguing video, or a hands-on exhibit—or, importantly, some combination of such experiences—has a cascading effect on learning choices and motivations, especially over the span of years (or even decades). Biographical studies of scientists and everyday citizens suggest that out-of-school STEM learning experiences can play a powerful role in shaping an individual pursuit of STEM careers or hobbies.[111] A broad range of evaluation approaches that capture the complexity of STEM learning and interest development across time and settings are needed in order to better understand how young people make connections across settings and experiences, and what elements of those connections contribute to the continuities that support sustained engagement and learning.

Change at Multiple Levels

As noted throughout this report, an individual is not the only point of change or growth in a STEM learning system. Communities, organizations, programs, and small groups (peers, friends, and families) undergo changes and transformations over time, moving to new ways of thinking and doing.[112] A group, program, organization, or community may change its objectives, organizational structure, resource allocation, established policies and procedures, styles of interaction, levels of collegiality, and even membership in pursuit of greater effectiveness, efficiency, or enjoyment, among other goals. This growth can be in terms of understanding of or interest in STEM, just as with individuals, and such growth is an important object of analysis for evaluators.[113]

Idiosyncratic Nature of Learning

A critical issue in evaluating out-of-school STEM programs is that learning occurs in diverse and unpredictable ways. For example, ethnographic studies of children's engagement in science outside of school,[114] and retrospective studies of scientists, science teachers, and science-interested individuals show that there are multiple pathways to developing enduring interests among young people.[115] Examining the scientific talk of young people makes clear that their personal feelings, intentions, purposes, and preferences shape their forms of engagement and ideas. It is also clear that talk-focused studies typically prioritize Western middle-class forms of talk as evidence of understanding.

Evaluators' awareness of the idiosyncratic nature of learning is important for ensuring that indicators and measures are not exclusively focused on predetermined outcomes and dominant social norms. An openness to and investigation of unintended effects of a program or experience is important for ensuring that an evaluation does not prioritize easily measurable outcomes, which can contribute to narrowing the role of out-of-school STEM environments and the possibilities they offer. It is also essential that evaluators understand the cultural patterns of social discourse of participating communities so that evaluations accurately capture a program's effects.

Additional Evaluation Challenges

There are many additional challenges to evaluating STEM learning in out-of-school programs. Importantly, young people participate in out-of-school programs based on their interests and motivations and use program resources in different ways. Because of this, out-of-school program evaluators have little control over who participates in a program, which can make it difficult to know whether the outcomes of the evaluation could be replicated with different participants. In addition, if the differences in program experiences among the participants are not well understood, it is difficult to describe what led to any measurable outcomes. For example, young people who consistently attend an out-of-school program are more likely to reap the benefits, compared with those who attend sporadically.[116]

Understanding the key features of any STEM learning environment and being able to capture, categorize, and analyze participants' diverse responses are fundamental challenges in making sense of how such environments do and do not promote growth and change.

In the social context of most out-of-school settings, individual assessments, such as tests and surveys, would typically interrupt the normal flow of activity, not be expected by the participants, and negatively change the nature of the environment. For this reason, some evaluators working in the informal STEM field have been particularly concerned with developing unobtrusive means to measure and document learning in such settings. Unobtrusive assessments would be built into the learning experience—embedded in activities such as games or challenges—or be derived observationally from the natural interactions of participants. Such "naturalistic" assessment would rely on documenting the ways in which learners seek help, share ideas, notice one another's capabilities, build reputations, and in other ways notice and make use of resources in their environment.[117]

Evaluations of out-of-school programs typically document short-term outcomes. Since learning is understood to occur over time and across settings, it is important to take more comprehensive and layered approaches to evaluation by considering both short-term and long-term factors and outcomes. For example, an early evaluation might focus on short-term outcomes such as whether program goals were achieved and how the design of the program did or did not contribute to achieving those goals. The evaluation might also focus on how a given program fits within the larger learning ecosystem, documenting how it diversifies, deepens, or enhances possibilities for STEM learning in a given community. In addition, the evaluation could measure the consequences of individual differences among participants and longer-term outcomes.

A Three-Level Approach to Evaluating the STEM Learning Ecosystem

From an ecosystem perspective on learning, a comprehensive out-of-school STEM program evaluation includes measurement at three interrelated levels: individual, program, and community.

At the individual level, evaluation of the quality of an out-of-school program would include measures of an individual's intellectual development in STEM; positive STEM identity and dispositional development; and expansion of an individual's horizons (awareness, connections, and choices), in the context of life-long, academic, and career engagement with STEM. Measurement at the individual level, especially when conducted longitudinally, can shed light on how out-of-school programs are, individually and collectively, responsive to an individual's learning needs, perceptions of ability, and interest in STEM.

At the program level, evaluation can document the resources and opportunities provided by the out-of-school STEM program. Evaluation at this level can suggest the ways in which program design and implementation can be augmented to better support young people's intellectual and social and emotional engagement, and how responsive the program is to participants' interests and experiences. Program-level evaluation can also measure how a program intentionally engages participants with community resources and possibilities to expand horizons of project participants. Questions can be asked about the capacity of adult facilitators/educators and whether they have opportunities to enhance their skills.

Program-level evaluation that considers the dimensions of engagement in STEM, responsiveness to young people, and connectivity with community, would include descriptions of program activities, information on staff training and development, information about levels of participation, and if the participants also participate in other STEM learning experiences at school or in the community. Program-level evaluation allows staff and evaluators to connect a program's resources and activities with individual outcomes in order to see what is working well, for whom, and to consider opportunities for change. In addition, program-level evaluation allows staff and evaluators to connect programmatic resources and activities with community-level resources and activities.

At the community level, evaluation can focus on the distribution of diverse STEM learning opportunities (across domains, practices, and levels of advancement); the ways in which a given program is synergistic with the resources within a community and across settings; and the ways in which a program affects the community by expanding learning opportunities and brokering additional engagement in STEM learning across different community settings. Community-level indicators signal the extent to which community-level resources are in place to support effective out-of-school STEM programming, to support connections among in-school and out-of-school learning, and to identify any need for action.

An evaluation at the community level can inform program design. Asset mapping and needs analysis are fundamental to the design of both individual programs and a set of opportunities across a community. They can identify areas in which needs exist in a community and allow stakeholders to understand the nature of local opportunities, what may or may not be working well, and where to best invest resources and new design and implementation efforts. Such mapping work is ongoing in

Examples of Evaluation at the Individual, Program, and Community Levels

Individual-Level Evaluation The Detroit Area Pre-College Engineering Program (DAPCEP) is a nonprofit organization that involves partnerships with universities, training programs, and K-12 school systems to connect historically underrepresented youths with high-quality STEM learning experiences. DAPCEP engages youths in out-of-school STEM learning experiences across several years, providing hands-on mathematics and engineering activities. Program activities are led by classroom teachers who seek to make explicit connections between the content of the hands-on activities and the mathematics that youths work with at school. Students also engage with local industries and professionals to see how mathematics and engineering translate to jobs and careers. Program evaluation has documented the positive effects of DAPCEP on participants, on student high school graduation rates, college enrollment, and selection of STEM-related majors.*

Program-Level Evaluation Intel the Computer Clubhouse Network (ICCN)** uses program-level evaluation to inform programmatic decisions. The ICCN has long engaged evaluators to help analyze and document ways in which its approaches are shaping and affecting the lives of participating youths. Evaluation partners have conducted interviews, surveys, observations, and reviews of staff reports to both provide feedback to the organization and support its program development.

Community-Level Evaluation To better support the development and coordination of its ecosystem of STEM learning opportunities, the Mozilla Hive NYC Learning Network works with Hive NYC partners and the Hive Research Lab to capture and share best practices and collective wisdom. For example, in 2014, Hive NYC members and stakeholders convened meetings to develop principles and guidelines for "working open"—a model for reflective, evaluative practice to support the continuous improvement of programs and outcomes at the community level. The model includes rapid prototyping,*** public storytelling to illustrate key findings, community contributions for co-development of approaches, and making the content of the network's activities openly accessible.

---

*Bevan, B., Michalchik, V., Remold, J., Bhanot, R., and Shields, P. (2013). *Final Report of the Learning and Youth Research and Evaluation Center.* San Francisco, CA: The Exploratorium.

**See http://www.computerclubhouse.org/ [May 2015].

***Rapid prototyping is the process of quickly fabricating a scale model using three-dimensional computer-aided design data.

the 42 statewide after-school networks[r] that have developed online repositories of STEM out-of-school curricula and information. The networks have also created an online database that maps STEM programming and connected learning opportunities. In addition, there are a number of guides to developing asset maps, including the W.K. Kellogg Foundation, the Community Tool Box, and Community Science.[s]

A three-level model would include evaluation of how the outcomes for individual participants are directly influenced by the program qualities and how both are shaped and supported by the community context. Evaluations of an out-of-school STEM program would focus on these elements, characteristics, and outcomes, while at the same time identifying any shortcomings, misalignments, and unintended effects, as well as any possibilities for new directions and innovations.

Work is now under way to develop new models for evaluations that may prove to be less disruptive, less obtrusive, and more meaningful than many commonly used near-term measures of individual learning and changes in attitudes and interests, such as surveys. One suggested approach is for evaluators to develop a framework for how formative (process), summative (outcome), and comparative evaluation interact. Existing measures[t] and program evaluations[u] are typically conducted at the individual or program level and typically focus on the short-term outcomes. Two notable exceptions are the longitudinal evaluations of 4-H Science[v] and FIRST®[w] (For Inspiration

[r]For example, see http://www.indianaafterschool.org/state/mapping-database/ [May 2015].

[s]For more information, see http://www.abcdinstitute.org/docs/kelloggabcd.pdf_[May 2015], http://ctb.ku.edu/en/table-of-contents [May 2015], and http://communityscience.com/knowledge4equity/AssetMappingToolkit.pdf [May 2015].

[t]For example, see the database of Assessment Tools for Informal Science (ATIS) at http://www.pearweb.org/atis/tools/browse?content=true [May 2015].

[u]For example, see the evaluations of public education programs at http://informalscience.org/evaluation/browse?type=evaluations [May 2015].

[v]For more information, see http://www.4-h.org/about/youth-development-research/positive-youth-development-study/ [August 2015].

[w]For more information, see http://www.usfirst.org/aboutus/impact [August 2015].

and Recognition of Science and Technology).[x] Greater investments into developing methods for longitudinal and community-level evaluations would make it possible for more evaluations to take an ecosystem perspective.

For out-of-school programs, for example, immediate measures of individual experiences could be developed to provide formative feedback to program leaders. Such measures could include what individuals are interested in or confused about. The resulting data could be used for program design and implementation. Long-term outcome measures, such as levels of interest in STEM or documentation of course and career choices, could be used to evaluate whether a program achieved its targeted goals and outcomes, and how it did so. Similarly, program measures could be seen as formative from the community perspective by addressing such questions as: Where are investments needed? Where are opportunities for action? What community resources might strengthen a program?

Common Instrumentation

The need to both consolidate and diversify evaluation methods at the individual level is an active area of research. Some researchers have pursued the development of standard metrics for measuring STEM interest and motivation across the range of out-of-school STEM environments.[118] Others use qualitative means to probe and document the way that out-of-school experiences shape young people's life trajectories, as evidenced by choices, pathways, and "ways of being"—for example, interacting with phenomena or appraising ideas, designs, and products.[119]

The most common approaches to research and evaluation focus on near-term measures that are easy to administer and score. Well-designed tools of this kind are an important component of an evaluation toolkit, and there are several ambitious initiatives under way to develop suites of tools that can be shared across projects:

- the Youth, Engagement, Attitudes and Knowledge (YEAK) Survey developed by 4-H;[y]
- the suite of tools developed by the Program in Education Afterschool and Resiliency (PEAR) at Harvard University;[z]
- the measures developed by the Activation Lab, a collaboration among the Learning Research and Development Center at the University of Pittsburgh, the Lawrence Hall of Science at University of California, Berkeley, and SRI International;[aa] and

[x]Two exceptions to the general statement about existing measures and program evaluations were added to the text.
[y]For more information, see http://www.4-h.org/about/youth-development-research/science-program-research/ [May 2015].
[z]For more information, see http://www.pearweb.org/tools/STEM.html [May 2015].
[aa]For more information, see http://www.activationlab.org/tools/ [May 2015].

- the Developing, Validating, and Implementing Standardized Evaluation Instruments (DEVISE) Project at Cornell University.[ab]

Policy makers understandably want a single, low-cost, easy-to-administer tool that can provide data that allow them to measure the effects of educational investments. Creating a single metric that could be used in the diversity of out-of-school STEM programs will not be simple[120] because it needs to be sensitive to differences among individuals (e.g., age, culture, level of participation) and programs (e.g., intensity and length, delivery method, goals) while not intruding on the program's design. Yet progress has been made in developing some common standardized measures that can track the long-term trajectories of young people's development and possibly (if linked to detailed accounting of program and community arrangements) also provide understanding across programs as to what elements of out-of-school settings and programs contribute to learning. Such measurement instruments allow for the comparison and aggregation of data across programs and settings. However, there are significant concerns about the ways that common measurements may constrict educational opportunities and approaches in schools and otherwise negatively affect learning in out-of-school settings. Clearly there are benefits and limitations of common metrics, and this is an area of work that deserves careful investment and study over time.[121]

Common approaches to measurement of youth outcomes are generally meant to document the contributions of out-of-school programs to STEM learning[122] or to determine whether the contributions to STEM learning vary for different populations of young people. Since common metrics are used for these purposes, evaluators need to continue to gauge whether program goals are being accomplished and whether there are any unintended consequences (e.g., intruding on program designs, using any one measure as the sole metric of outcomes).

The efforts to develop common metrics of important constructs—such as learning, engagement, and identity—have generated conversations about what should "count" as outcomes of out-of-school STEM programs and for which outcomes out-of-school programs should be held accountable. Work in this area has included metrics for measuring both youth outcomes[ac] and program quality.[ad]

[ab]For more information, see http://www.birds.cornell.edu/page.aspx?pid=1677 [May 2015].

[ac]See the Common Instrument at http://www.pearweb.org/tools/commoninstrument.html [May 2015] and the Youth, Engagement, Attitudes and Knowledge Survey at http://www.4-h.org/about/youth-development-research/science-program-research/ [May 2015].

[ad]See the Dimensions of Success observation tool at http://www.pearweb.org/tools/dos.html [May 2015] and the Youth Program Quality Assessment at http://www.cypq.org/assessment [May 2015].

WHAT IS KNOWN AND RECOMMENDATIONS FOR ACTION

As the number and diversity of out-of-school programs that support STEM learning continue to grow, it is becoming increasingly important for policy makers and funders to make informed decisions about which programs to support. The existing research provides important information that can help inform some of those difficult decisions. The committee offers three criteria for identifying and developing productive out-of-school STEM learning programs, and six recommendations for actions that policy makers, educators, and other stakeholders can take to support programs that reflect the criteria.

Criteria For Productive Programs

Young people (ages 5-18) develop an understanding of STEM concepts and skills through an iterative process across a wide array of learning experiences that take place in both out-of-school programs and in school.[123] The iterative process of learning STEM requires policy makers to create funding streams and policies that encourage productive out-of-school STEM experiences and how to link them in order to create sets of coherent learning opportunities. Policy makers at the local, state, and national levels have different mechanisms available to them for achieving these goals, and they can each play a role in supporting such education reform.

Although opportunities to engage in STEM activities in out-of-school settings and programs is sometimes thought of as an optional enrichment opportunity, this perspective is not consistent with what is known about the outcomes of such settings and programs. Access to productive out-of-school opportunities that engage young people in authentic STEM experiences is a critical piece of the STEM learning ecosystem. Such out-of-school opportunities can support STEM learning independently from classroom learning, and they are particularly well suited to building interest in STEM and identity as a STEM learner.[124]

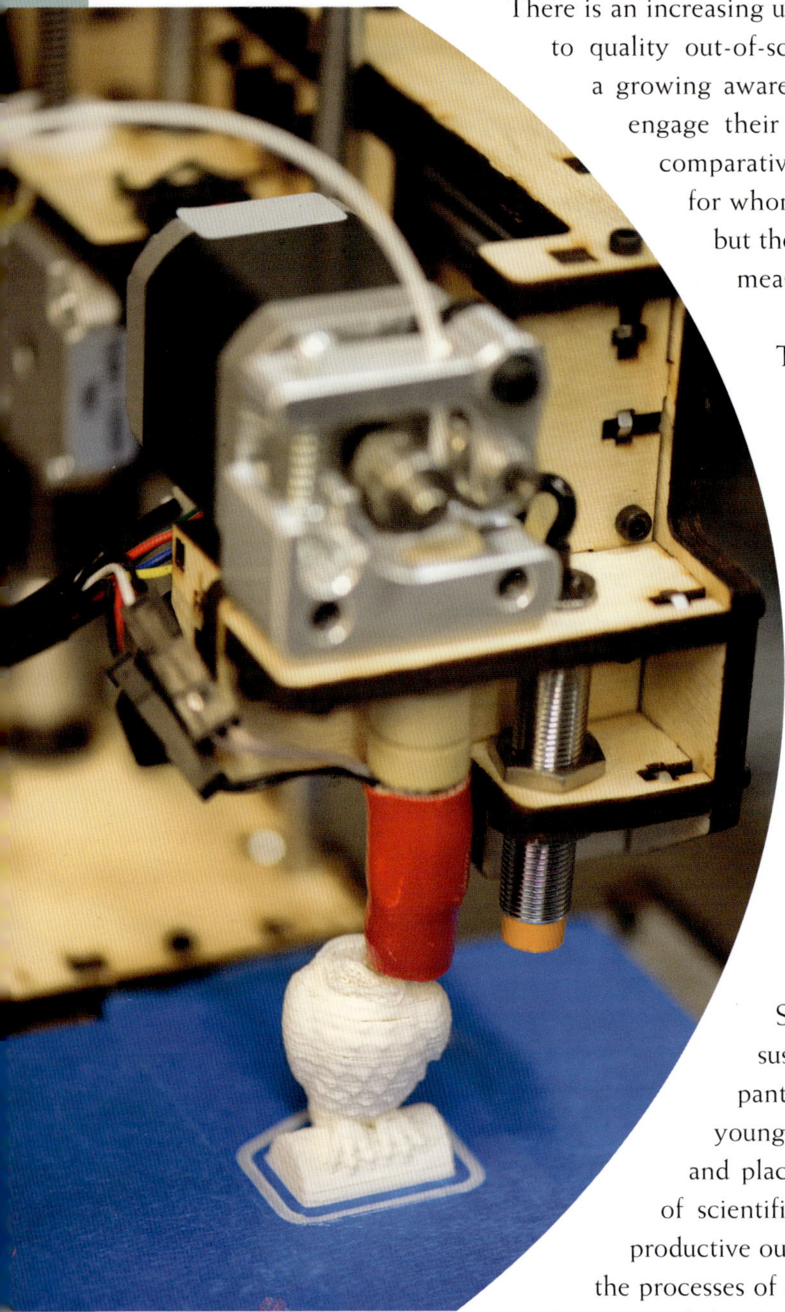

There is an increasing understanding of how to broaden and deepen access to quality out-of-school programs that support STEM learning and a growing awareness of the need to make it easier for families to engage their children.[125] Clear evidence from summative and comparative program evaluations of what programs work best for whom and under what circumstances does not yet exist, but the field is taking steps to develop new and meaningful measurement strategies.

To support informed policy and program decision making we concluded that there are three criteria for identifying and developing productive out-of-school STEM learning programs. Together, the criteria represent the ways in which youth development, STEM learning in informal environments, and learning across settings intertwine to support productive STEM out-of-school programs that successfully engage young people in STEM learning and actively support inclusion and broaden participation by young people in STEM learning.

1. **Productive programs engage young people intellectually, socially, and emotionally.** Productive out-of-school STEM programs provide young people with first-hand experiences with STEM phenomena and materials, engage them in sustained STEM practices, and are aligned with participants' cultural resources and practices. In such programs, young people are engaged in first-hand, materials-rich, and place-based learning experiences that involve processes of scientific or engineering investigation and practice. Thus, productive out-of-school STEM programs engage young people in the processes of doing STEM in ways they find compelling and challenging and develop their interest, understanding, and commitment to continue engaging in STEM learning.

2. **Productive programs respond to young people's interests, experiences, and cultural practices.** Productive out-of-school STEM programs make STEM relevant to the questions that interest young people, support collaboration and leadership by young people, and train

staff to support and build young people's STEM activities and interest. Productive out-of-school STEM programs are also responsive to young people's prior interests and experiences so that they can see STEM as meaningful and relevant to their own experiences and aspirations.

3. **Productive programs connect STEM learning in out-of-school, school, home, and other settings.** Productive out-of-school STEM programs explicitly help young people make connections among STEM experiences in and across settings and programs, leveraging community resources and partnerships, and brokering ongoing opportunities to engage in STEM learning activities. Productive out-of-school programs also help young people to understand how what they experience and learn relates to learning in other settings, including school. Thus, productive out-of-school programs purposefully help young people, their parents, and others in the community capitalize on developing expertise and interests across time and setting.

Evaluating Programs and Generating New Evidence

Generating evidence of productive out-of-school STEM programs is conceptually and practically complicated by the fact that STEM learning accumulates over time and across settings, change occurs at multiple levels (individual, program, and community), STEM learning is idiosyncratic, and the norms of out-of-school programs lead to practical barriers in administering assessments. There is a need for new conceptual tools and approaches to evaluation that can help generate hypotheses and theoretical accounts of STEM learning in out-of-school programs. Some of this work has begun: for example, there are efforts to develop common metrics and instruments to compare individual outcomes across a large number of programs, and there are efforts under way to develop innovative, unobtrusive approaches that are culturally responsive and honor the multiplicity of out-of-school program goals.

Next Steps

Policy makers, funders, and program leaders need to work together to sustain and expand a robust and iterative ecosystem of learning opportunities in schools and in out-of-school programs. We identified six actions that policy makers, program developers, and other stakeholders should take to support programs that reflect the criteria for identifying and designing productive out-of-school STEM learning programs:

1. Understand the local conditions for creating an ecosystem of high-quality productive out-of-school STEM learning programs: **Build a map and bridge the gaps.**
2. Design programs to achieve access, equity, continuity, and coherence: **Connect young people with opportunities to learn.**

3. Support the use of creative and responsive approaches to evaluating the success of programs at the individual, program, and community levels: **Support innovative evaluation approaches.**

4. Increase the professionalization of out-of-school program leaders and staff: **Provide professional development.**

5. Strengthen the STEM learning infrastructure: **Build an infrastructure that will last.**

6. Invest in research to improve our understanding of STEM learning in out-of-school programs: **Explore how STEM learning ecosystems work.**

The rest of this chapter elaborates on these important next steps.

BUILD A MAP AND BRIDGE THE GAPS

Mapping existing STEM learning resources and gaps is a critical first step in supporting a robust STEM learning ecosystem that can meet the interests and needs of all young people through a wide variety of intellectually compelling and culturally responsive programs.

Every community has a unique set of learning resources available to young people: they include natural settings, industries, universities, and local community-based and youth development organizations. As discussed in the second chapter of this report, productive programs provide compelling, responsive, and connected learning experiences in STEM.

To ensure that a wide variety of developmentally appropriate opportunities in a STEM learning ecosystem are available to all, there is a need for educational leaders to inventory existing resources, as well as gaps in opportunities, that both reinforce and expand on opportunities in schools. The resulting regional or community STEM learning map should guide program investments and help identify opportunities to leverage existing resources and experiences. Funders and policy makers should encourage program leaders to develop or review existing STEM learning maps to increase the potential for return on their investments and identify opportunities for partnerships.

CONNECT YOUNG PEOPLE WITH OPPORTUNITIES TO LEARN

To support equitable access and participation in out-of-school opportunities to learn STEM, there is a need to identify and train brokers or develop brokering mechanisms that can help families and young people, especially from groups historically underrepresented in STEM, to identify and access settings and programs that help young people "take the next step" in their STEM learning.

A robust STEM learning ecosystem is only effective in the long run if its many and varied opportunities are apparent and available to all school-age children in the community. Funders and policy makers should support efforts to develop brokers who can connect young people with STEM learning opportunities.

Creating connections among learning opportunities will require program managers to provide connections to other programs and opportunities for learning. As young people's interests deepen or shift, adults need to identify and direct them to new programs or opportunities in which they can advance their learning and pursue out-of-school STEM experiences. Communities need brokers who understand the interests and needs of the young people in their communities and of the STEM learning opportunities available. Brokers can benefit from participation in regional networks that include other brokers and program leaders to enrich and connect opportunities in their communities.

Support Innovative Evaluation Approaches

To evaluate out-of-school programs, the field needs innovative measures for program evaluation that will not impinge on the nature of out-of-school learning experiences, are culturally responsive, and are flexible enough to address a wide range of program goals.

A robust STEM learning ecosystem offers a wide variety of programs and opportunities that meet the varied needs of young people and has positive effects on individuals, programs, and communities. To better understand this ecosystem, education leaders, funders, and policy makers should support the development of innovative evaluation approaches that are valid in out-of-school STEM environments, are locally and culturally responsive, and honor the multiplicity of program goals. From an ecosystem perspective, measures need to take into account how young people learn over time; thus, longitudinal studies and innovations in assessment that account for development over time are essential.

A central principle for such novel approaches should be that they do not inadvertently formalize informal settings or disrupt young people's learning experiences. In addition, there is a need for evaluations that yield rich descriptions of community contexts, program implementation, and learner experiences. Innovations from other fields, such as youth development, should be brought to the out-of-school STEM ecosystem to better investigate the characteristics and qualities of programs.

Provide Professional Development

To support productive and responsive teaching and learning in out-of-school settings and programs, program staff need opportunities to develop their ability to nurture young people's interests and understanding of STEM content and practices.

The variety of out-of-school settings and programs that support STEM learning is facilitated by educators and other adults who come to the field with a wide array of prior experiences. Education leaders and program managers should support the professional activities of program staff by planning for and providing ongoing opportunities for professional reflection and learning in content, pedagogy, and instructional design. Professional development should integrate research and practice from multiple disciplines, such as formal education, social work, developmental psychology, urban studies, and similar fields. It is important that time and compensation for participating in professional development activities are provided. Policy makers and funders should invest in efforts to create entry-level and ongoing professional development mechanisms for staff of out-of-school programs.

Build an Infrastructure That Will Last

To develop an effective, sustainable infrastructure of STEM in out-of-school programs for all young people, funders, community leaders, and program leaders need to work together to identify areas for investment, expansion, or redirection.

Only a fraction of the need for programs outside of school is being met, and not all existing programs provide high-quality STEM learning opportunities. Programs are supported by a variety of funding sources, including volunteer organizations, private foundation grants, and local, state, and national agencies; some are fee for service and some are free. To sustain the high-quality programs that are available today and to gradually increase the nation's capacity to meet existing and future needs, funders, community leaders, and program leaders need to develop a sustainable infrastructure to support long-term growth.

Funder networks should facilitate sharing and collaboration across programs, including both in-school and out-of-school efforts. Community networks should provide administrative support in such areas as professional development, evaluation, assessment, and brokering of opportunities. Networks for program leaders should share strategies for program design, staff development, and documentation of program effects. In building these professional networks and infrastructure, it is critical that they do not lead to a narrowing of possibilities for young people.

Explore How STEM Learning Ecosystems Work

To expand research-based knowledge about productive strategies to support STEM learning in out-of-school settings and programs, there is a need to invest in research that documents both the learning that occurs in individual programs and also how STEM learning develops across settings and over time.

To build on existing knowledge, policy makers and funders should invest in local research-practice partnerships that combine the wisdom of practice and understanding of local conditions and young people with expertise in research and evaluation, while recognizing the challenges to implementation and sustainability of program improvements. The work that is needed includes longitudinal studies of youth trajectories in STEM learning, studies that relate program strategies to learner experiences and outcomes, studies of how brokering local STEM learning opportunities can broaden participation in STEM, and studies that examine how formal and informal STEM learning program designs can reinforce and enrich one another. Also needed is comparative research into questions of how the strategies of different out-of-school STEM program affect participants' experiences and outcomes and how community or regional contexts influence program implementation and quality.

cot(-x) = -cot(x)
4px = y2
Log ax 1 (1+h)
x2 / a2 - y2 / b2 = 1
P = C e rt
tan(-x) = -tan(x)
sec(-x) = sec(x)
e = mc2

NOTES

[1]Peter, N. (2002). *Outcomes and Research in Out-of-School Time Program Design*. Philadelphia, PA: Best Practices Institute.

[2]National Research Council. (2009). *Learning Science in Informal Environments: People, Places, and Pursuits*. Committee on Learning Science in Informal Environments, P. Bell, B. Lewenstein, A.W. Shouse, and M.A. Feder (Eds.). Board on Science Education, Center for Education, Division of Behavioral and Social Sciences and Education. Washington, DC: The National Academies Press.

[3]Shankar, M., and Kalil, T. (2013). *Leveraging Mental Muscle for Academic Excellence*. Washington, DC: Office of Science and Technology Policy, Executive Office of the President. Available: http://www.whitehouse.gov/blog/2013/06/28/leveraging-mental-muscle-academic-excellence [February 2015].

[4]Barron, B. (2006). Interest and self-sustained learning as catalyst of development: A learning ecologies perspective. *Human Development, 49*, 193-224.

Bell, P., Bricker, L.A., Tzou, C., Lee, T., and Van Horne, K. (2012). Engaging learners in scientific practices related to obtaining, evaluating, and communicating information. *The Science Teacher, 79*(8), 31-36.

Falk, J.H., and Dierking, L.D. (2010). The 95% solution: School is not where most Americans learn most of their science. *American Scientist, 98*, 486-493.

Falk, J.H., Dierking, L.D., Osborne, J., Wenger, M., Dawson, E.. and Wong, B. (2015). Analyzing science education in the UK: Taking a system-wide approach. *Science Education, 99*(1), 145-173.

Gutiérrez, K., and Rogoff, B. (2003). Cultural ways of learning: Individual traits or repertoires of practice. *Educational Researcher, 32*(5), 19-25.

Traphagen, K., and Traill, S. (2014). *How Cross-Sector Collaborations Are Advancing STEM Learning*. Palo Alto, CA: Noyce Foundation.

[5]Bronfenbrenner, U. (1977). Toward an experimental ecology of human development. *American Psychologist, 32*(7), 513-531.

Bronfenbrenner, U., and Morris, P.A. (2006). The bioecological model of human development. In W. Damon and R.M. Lerner (Eds.), *Handbook of Child Psychology, Volume 1: Theoretical Models of Human Development* (6th ed., pp. 793-828). New York: Wiley.

Krishnamurthi, A. (2015). *STEM Learning Across Settings: Cultivating Learning Ecosystems*. Afterschool Advocate Newsletter-Afterschool Snack. Available: http://www.afterschoolalliance.org/afterschoolsnack/ASnack.cfm?idBlog=42F434BF-215A-A6B3-02FE5A2917C-C75A9 [January 2015].

McLaughlin, M. (2000). *Community Counts: How Youth Organizations Matter for Youth Development*. Washington, DC: Public Education Network.

[6]Bevan, B., with Dillon, J., Hein, G.E., Macdonald, M., Michalchik, V., Miller, D., Root, D., Rudder, L., Xanthoudaki, M., and Yoon, S. (2010). *Making Science Matter: Collaborations Between Informal Science Education Organizations and Schools. A CAISE Inquiry Group Report*. Washington, DC: Center for Advancement of Informal Science Education.

National Research Council. (2009). *Learning Science in Informal Environments: People, Places, and Pursuits*. Committee on Learning Science in Informal Environments. P. Bell, B. Lewenstein, A.W. Shouse, and M.A. Feder (Eds.). Board on Science Education, Center for Education. Division of Behavioral and Social Sciences and Education. Washington, DC: The National Academies Press.

Traphagen, K., and Traill, S. (2014). *How Cross-Sector Collaborations Are Advancing STEM Learning*. Palo Alto, CA: Noyce Foundation.

[7]Overton, W.F. (2015). Processes, relations, and relational-developmental-systems. In W.F. Overton and P.C.M. Molenaar (Eds.), *Theory and Method: Handbook of Child Psychology and Developmental Science, Volume 1, Seventh Edition* (pp. 9-62). Hoboken, NJ: Wiley.

[8]Elder, G.H., Shanahan, M.J., and Jennings, J. (2015). Human development in time and place. In R. Lerner, M. Bornstein, and T. Levanthal (Eds.), *Handbook of Child Psychology and Developmental Science: Ecological Settings and Processes in Developmental Systems* (pp. 1-49). New York: Wiley.

[9]National Research Council. (2011). *Successful K-12 STEM Education: Identifying Effective Approaches in Science, Technology, Engineering, and Mathematics*. Committee on Highly Successful Science Programs for K-12 Science Education. Board on Science Education and Board on Testing and Assessment, Division of Behavioral and Social Sciences and Education. Washington, DC: The National Academies Press.

[10]Jolly, E., Campbell, P., and Perlman, L. (2004). *Engagement, Capacity, and Continuity: A Trilogy for Student Success*. Groton, MA, and Minneapolis, MN: Campbell-Kibler Associates and Science Museum of Minnesota.

National Research Council. (2011). *Successful K-12 STEM Education: Identifying Effective Approaches in Science, Technology, Engineering, and Mathematics*. Committee on Highly Successful Science Programs for K-12 Science Education. Board on Science Education and Board on Testing and Assessment, Division of Behavioral and Social Sciences and Education. Washington, DC: The National Academies Press.

[11]Stevens, R., Bransford, J., and Stevens, A. (2005). *The LIFE Center's Life-long and Life-wide Diagram*. Available: http://life-slc.org/about/citationdetails.html [February 2015].

[12]Dierking, L.D., Falk, J.H., Holland, G., Fisher, S., Schatz, D., and Wilke, L. (1997). *Collaboration: Critical Criteria for Success*. Washington, DC: Association of Science-Technology Centers.

Lee, J.J., and Hammer, J. (2011). Gamification in education: What, how, why bother? *Academic Exchange Quarterly, 15*(2), 1-5.

Hill, N.E., Tyson, D.F., and Bromell, L. (2009). Parental involvement in middle school: Developmentally appropriate strategies across SES and ethnicity. In N.E. Hill and R.K. Chao (Eds.), *Families, Schools, and the Adolescent: Connecting Research, Policy, and Practice* (pp. 53-72). New York: Teachers College Press.

National Research Council. (2009). *Learning Science in Informal Environments: People, Places, and Pursuits*. Committee on Learning Science in Informal Environments. P. Bell, B. Lewenstein, A.W. Shouse, and M.A. Feder (Eds.). Board on Science Education, Center for Education. Division of Behavioral and Social Sciences and Education. Washington, DC: The National Academies Press.

[13]Afterschool Alliance. (2013). *Defining Youth Outcomes for STEM Learning in Afterschool*. Available: http://www.afterschoolalliance.org/STEM_Outcomes_2013.pdf [February 2015].

National Research Council. (2009). *Learning Science in Informal Environments: People, Places, and Pursuits*. Committee on Learning Science in Informal Environments. P. Bell, B. Lewenstein, A.W. Shouse, and M.A. Feder (Eds.). Board on Science Education, Center for Education, Division of Behavioral and Social Sciences and Education. Washington, DC: The National Academies Press.

[14]National Research Council. (2009). *Learning Science in Informal Environments: People, Places, and Pursuits*. Committee on Learning Science in Informal Environments. P. Bell, B. Lewenstein, A.W. Shouse, and M.A. Feder (Eds.). Board on Science Education, Center for Education, Division of Behavioral and Social Sciences and Education. Washington, DC: The National Academies Press.

[15]Ibid.

[16]U.S. Department of Education and U.S. Department of Justice. (2000). *Working for Children and Families: Safe and Smart Afterschool Programs*. Available: http://www2.ed.gov/offices/OESE/archives/pubs/parents/SafeSmart/green-1.doc [February 2015].

[17]National Research Council. (2009). *Learning Science in Informal Environments: People, Places, and Pursuits*. Committee on Learning Science in Informal Environments. P. Bell, B. Lewenstein, A.W. Shouse, and M.A. Feder (Eds.). Board on Science Education, Center for Education, Division of Behavioral and Social Sciences and Education. Washington, DC: The National Academies Press.

[18]Pierce, K.M., Auger, A., and Vandell, D.L. (2013). *Narrowing the Achievement Gap: Consistency and Intensity of Structured Activities During Elementary School*. Paper presented at the Society for Research in Child Development Biennial Meeting, April, Seattle, WA.

Vandell, D.L., Reisner, R.R., and Pierce, K.M. (2007). *Outcomes Linked to High-Quality Afterschool Programs: A Longitudinal Study of Promising Afterschool Programs*. Unpublished report prepared for the Charles Stewart Mott Foundation. Available: http://www.education.uci.edu/childcare/pdf/afterschool/PP%20Longitudinal%20Findings%20Final%20Report.pdf [February 2015].

[19]Christensen, R., Knezek, G., and Tyler-Wood, T. (2015). A retrospective analysis of STEM career interest among mathematics and science academy students. *International Journal of Learning, Teaching, and Educational Research, 10*(1), 45-58.

Garg, R., Kauppi, C., Urajnik, D., and Lewko, J. (2007). A longitudinal study of the effects of context and experience on the scientific career choices of Canadian adolescents. *Canadian Journal of Career Development, 9*(1), 15-24. Available: http://ceric.ca/cjcd/archives/v9-n1/article2.pdf [February 2015].

Jones, G., Taylor, A., and Forrester, J.H. (2011). Developing a scientist: A retrospective look. *International Journal of Science Education, 33*(12), 1653-1673.

Tan, E., and Calabrese Barton, A. (2007). From peripheral to central, the story of Melanie's metamorphosis in an urban middle school science class. *Science Education, 92*(4), 567-590.

[20]Bell, P., Bricker, L., Reeve, S., Toomey Zimmerman, H., and Tzou, C. (2013). Discovering and supporting successful learning pathways of youth in and out of school: Accounting for the development of everyday expertise across settings. In B. Bevan, P. Bell, R. Stevens, and A. Razfar (Eds.), *LOST Opportunities: Learning in Out-of-School Time* (pp. 119-140). Heidelberg, Germany: Springer Netherlands.

[21]Afterschool Alliance. (2014). *America After 3 PM: Afterschool Programs in Demand*. Washington, DC: Author. Available: http://afterschoolalliance.org/documents/AA3PM-2014/AA3PM_National_Report.pdf [February 2015].

[22]Ibid.

[23]21st Century Community Learning Centers Program, Office of Academic Improvement Programs, Office of Elementary and Secondary Education, U.S. Department of Education. (2015). *Programs, 21st Century Community Learning Centers home page*. Available: http://www2.ed.gov/programs/21stcclc/awards.html [February 2015].

[24]Afterschool Alliance. (2014). *America After 3 PM: Afterschool Programs in Demand*. Washington, DC: Author. Available: http://afterschoolalliance.org/documents/AA3PM-2014/AA3PM_National_Report.pdf [February 2015].

[25]House, A., Llorente, C., Leones, T., and Lundh, P. (2014). *Navigating the Future of Afterschool Science: Afterschool Science Networks Study Recommendations*. Menlo Park, CA: SRI International.

[26]National Research Council and Institute of Medicine. (2002). *Community Programs to Promote Youth Development*. Committee on Community-Level Programs for Youth. J. Eccles and J.A. Gootman (Eds.). Board on Children, Youth, and Families, Division of Behavioral and Social Sciences and Education. Washington, DC: National Academy Press.

[27]National Research Council and Institute of Medicine. (2002). *Community Programs to Promote Youth Development*. Committee on Community-Level Programs for Youth. J. Eccles and J.A. Gootman (Eds.). Board on Children, Youth, and Families, Division of Behavioral and Social Sciences and Education. Washington, DC: National Academy Press.

[28]National Research Council. (2009). *Learning Science in Informal Environments: People, Places, and Pursuits*. Committee on Learning Science in Informal Environments. P. Bell, B. Lewenstein, A.W. Shouse, and M.A. Feder (Eds.). Board on Science Education, Center for Education, Division of Behavioral and Social Sciences and Education. Washington, DC: The National Academies Press.

[29]Banks, J.A., Au, K.H., Ball, A.F., Bell, P., Gordon, E.W., Gutiérrez, K.D., and Zhou, M. (2007). *Learning in and out of School in Diverse Environments: Life-long, Life-wide, Life-deep*. Seattle, WA: The LIFE Center, University of Washington, Stanford University, and SRI International and Center for Multicultural Education, University of Washington. Available: http://life-slc.org/docs/Banks_etal-LIFE-Diversity-Report.pdf [May 2015].

Ito, M., Gutiérrez, K., Livingstone, S., Penuel, W., Rhodes, J., Salen, K., Schor, J., Sefton-Green, J., and Watkins, S.C. (2012). *Connected Learning: An Agenda for Research and Design*. Chicago, IL: MacArthur Foundation.

National Academy of Engineering and National Research Council. (2014). *STEM Integration in K-12 Education: Status, Prospects, and an Agenda for Research*. M. Honey, G. Pearson, and H. Schweinguber (Eds.). Committee on Integrated STEM Education. Washington, DC: The National Academies Press.

[30]Gutiérrez, K., and Vossoughi, S. (2010). "Lifting off the ground to return anew": Documenting and designing for equity and transformation through social design experiments. *Journal of Teacher Education, 61*(1-2), 100-117.

Nasir, N., Rosebery, A., Warren, B., and Lee, C.D. (2006). Learning as a cultural process: Achieving equity through diversity. In K. Sawyer (Ed.), *The Cambridge Handbook of the Learning Sciences* (pp. 489-504). Cambridge, UK: Cambridge University Press.

National Research Council and Institute of Medicine. (2002). *Community Programs to Promote Youth Development*. Committee on Community-Level Programs for Youth. J. Eccles, and J.A. Gootman (Eds.). Board on Children, Youth, and Families, Division of Behavioral and Social Sciences and Education. Washington, DC: National Academy Press.

[31]Falk, J.H., Dierking, L.D., Staus, N., Penuel, W., Wyld, J., and Bailey, D. (in press). Understanding and connecting youth STEM interest and participation across the community: The SYNERGIES Project. *International Journal of Science Education, Part B.*

Ito, M., Gutiérrez, K., Livingstone, S., Penuel, W., Rhodes, J., Salen, K., Schor, J., Sefton-Green, J., and Watkins, S.C. (2012). *Connected Learning: An Agenda for Research and Design*. Chicago, IL: MacArthur Foundation.

National Research Council. (2009). *Learning Science in Informal Environments: People, Places, and Pursuits.* Committee on Learning Science in Informal Environments. P. Bell, B. Lewenstein, A.W. Shouse, and M.A. Feder (Eds.). Board on Science Education, Center for Education, Division of Behavioral and Social Sciences and Education. Washington, DC: The National Academies Press.

[32]Bang, M., Warren, B., Rosebery, A., and Medin, D. (2012). Desettling expectations in science education. *Human Development, 55*, 302-318.

Jolly, E., Campbell, P., and Perlman, L. (2004). *Engagement, Capacity, and Continuity: A Trilogy for Student Success*. Groton, MA, and Minneapolis: Campbell-Kibler Associates and Science Museum of Minnesota.

National Academy of Engineering and National Research Council. (2014). *STEM Integration in K-12 Education: Status, Prospects, and an Agenda for Research.* M. Honey, G. Pearson, and H. Schweinguber (Eds.). Committee on Integrated STEM Education. Washington, DC: The National Academies Press.

[33]National Research Council. (2011). *Successful K-12 STEM Education: Identifying Effective Approaches in Science, Technology, Engineering, and Mathematics*. Committee on Highly Successful Science Programs for K-12 Science Education. Board on Science Education and Board on Testing and Assessment, Division of Behavioral and Social Sciences and Education. Washington, DC: The National Academies Press.

House, A., Llorente, C., Leones, T., and Lundh, P. (2014). *Navigating the Future of Afterschool Science: Afterschool Science Networks Study Recommendations*. Menlo Park, CA: SRI International.

[34]Ibid.

[35]National Research Council. (2009). *Learning Science in Informal Environments: People, Places, and Pursuits.* Committee on Learning Science in Informal Environments. P. Bell, B. Lewenstein, A.W. Shouse, and M.A. Feder (Eds.). Board on Science Education, Center for Education, Division of Behavioral and Social Sciences and Education. Washington, DC: The National Academies Press.

Barton, A., and Roth, W. (2005). Rethinking scientific literacy. *Canadian Journal of Education, 28*(3), 561-566.

[36]Heath, S., and McLaughlin, M. (1994). The best of both worlds: Connecting schools and community youth organizations for all-day, all-year learning. *Educational Administration Quarterly, 30*(3), 278-300.

National Research Council and Institute of Medicine. (2002). *Community Programs to Promote Youth Development.* Committee on Community-Level Programs for Youth. J. Eccles and J.A. Gootman (Eds.). Board on Children, Youth, and Families, Division of Behavioral and Social Sciences and Education. Washington, DC: National Academy Press.

Mahoney, J.L., Lord, H., and Carryl, E. (2005). An ecological analysis of after-school program participation and the development of academic performance and motivational attributes for disadvantaged children. *Child Development, 76*, 811-825.

[37]Bowers, E.P., Li, Y., Kiely, M.K., Brittian, A., Lerner, J.V., and Lerner, R.M. (2010). The five Cs model of positive youth development: A longitudinal analysis of confirmatory factor structure and measurement invariance. *Journal of Youth and Adolescence, 39*(7), 720-735.

Jelicic, H., Bobek, D., Phelps, E.D., Lerner, J.V., and Lerner, R.M. (2007). Using positive youth development to predict contribution and risk behaviors in early adolescence: Findings from the first two waves of the 4-H Study of Positive Youth Development. *International Journal of Behavioral Development, 31*(3), 263-273.

Lerner, R.M., Lerner, J.V., Lewin-Bizan, S., Bowers, E.P., Boyd, M., Mueller, M., Schmid, K., and Napolitano, C. (2011). Positive youth development: Processes, programs, and problematics. *Journal of Youth Development, 6*(3), 40-64.

National Research Council and Institute of Medicine. (2002). *Community Programs to Promote Youth Development.* Committee on Community-Level Programs for Youth, J. Eccles, and J.A. Gootman (Eds.). Board on Children, Youth, and Families, Division of Behavioral and Social Sciences and Education. Washington, DC: National Academy Press.

Phelps, E., Zimmerman, S., Warren, A.E.A., Jelicic, H., von Eye, A., and Lerner, R.M. (2009). The structure and developmental course of positive youth development (PYD) in early adolescence: Implications for theory and practice. *Journal of Applied Developmental Psychology, 30*(5), 571-584.

[38]Bruner, J. (1960). *The Process of Education.* Cambridge, MA: Harvard University Press.

Bedderman, T. (1982). What research says: Activity science—the evidence shows it matters. *Science and Children, 20*(1), 39-41.

National Research Council. (2012). *A Framework for K-12 Science Education: Practices, Crosscutting Concepts, and Core Ideas*. Committee on Conceptual Framework for the New K-12 Science Education Standards. Board on Science Education, Division of Behavioral and Social Sciences and Education. Washington, DC: The National Academies Press.

[39]Remold, R., Verdugo, R., and Michalchik, V. (2013). *California Tinkering Afterschool Network Pilot Year Evaluation Report*. Menlo Park, CA: SRI International. Available: http://www.sri.com/work/publications/california-tinkering-afterschool-network [February 2015].

[40]Furtak, E., Seidel, T., Iverson, H., and Briggs, D.C. (2013). Experimental and quasi-experimental studies of inquiry-based science teaching: A meta-analysis. *Review of Educational Research, 82*(3), 300-329.

Minner, D.D., Levy, A.J., and Century, J. (2010). Inquiry based science instruction—What is it and does it matter? Results from a research synthesis years 1984 to 2002. *Journal of Research in Science Teaching, 47*(4), 474-496.

[41]Bedderman, T. (1982). What research says: Activity science—the evidence shows it matters. *Science and Children, 20*(1), 39-41.

[42]National Research Council. (2012). *A Framework for K-12 Science Education: Practices, Crosscutting Concepts, and Core Ideas*. Committee on Conceptual Framework for the New K-12 Science Education Standards. Board on Science Education, Division of Behavioral and Social Sciences and Education. Washington, DC: The National Academies Press.

[43]Ibid.

[44]Vossoughi, S., Escudé, M., Kong, F., and Hooper, P. (2013). *Tinkering, Learning, and Equity in the After-School Setting*. Paper presented at the FabLearn Conference, Stanford University, Stanford, CA.

[45]Granger, R., Durlak, J.A., Yohalem, N., and Reisner, E. (2007). *Improving After-School Program Quality*. New York: William T. Grant Foundation. Available: http://www.wtgrantfoundation.org/publications_and_reports/browse_reports/Imp_AS_Granger_Yohalem_Durlak [February 2015].

Mahoney, J.L., Parente, M.E., and Zigler, E.F. (2010). After-school program participation and children's development. In J. Meece and J.S. Eccles (Eds.), *Handbook of Research on Schools, Schooling, and Human Development* (pp. 379-397). New York: Routledge.

[46]Bevan, B., Michalchik, V., Bhanot, R., Rauch, N., Remold, J., Semper, R., and Shields, P. (2010). *Out-of-School Time STEM: Building Experience, Building Bridges. Trends Questions and Findings from the Field*. San Francisco: The Exploratorium. Available: http://stelar.edc.org/sites/stelar.edc.org/files/STEM_OST_Conf_Report.pdf [May 2015].

[47]Calabrese Barton, A., and Tan, E. (2009). Funds of knowledge and discourses and hybrid space. *Journal of Research in Science Teaching, 46*(1), 50-73.

Immordino-Yang, M.H., and Damasio, A.R. (2007). We feel, therefore we learn: The relevance of affective and social neuroscience to education. *Mind, Brain, and Education, 1*(1), 3-10.

[48]Lederman, N.G., Abd-El-Khalick, F., Bell, R.L., and Schwartz, R.S. (2002). Views of Nature of Science Questionnaire: Toward valid and meaningful assessment of learners' conceptions of nature of science. *Journal of Research in Science Teaching, 39*(6), 497-521.

[49]Finson, K.D. (2001). Applicability of the DAST-C to the images of scientists drawn by students of different racial groups. *Journal of Elementary Science Education, 15*, 15-26.

Finson, K.D. (2002). Drawing a scientist: What we do and do not after fifty years of drawings. *School Science and Mathematics, 102*, 335-346.

[50]Newton, D.P., and Newton, L.D. (1992). Young children's perceptions of science and the scientist. *International Journal of Science Education, 14*, 331-348.

Barman, C.R. (1996). How do students really view science and scientists? *Science and Children, 34*, 30-33.

Dweck, C.S. (2006). Is math a gift? Beliefs that put females at risk. In S.J. Ceci and W.M. Williams (Eds.), *Why Aren't More Women in Science? Top Researchers Debate the Evidence*. Washington, DC: American Psychological Association.

[51]Engle, R.A. (2006). Framing interactions to foster generative learning: A situative explanation of transfer in a community of learners classroom. *Journal of the Learning Sciences, 15*(4), 451-498.

National Research Council. (2012). *A Framework for K-12 Science Education: Practices, Crosscutting Concepts, and Core Ideas*. Committee on Conceptual Framework for the New K-12 Science Education Standards. Board on Science Education, Division of Behavioral and Social Sciences and Education. Washington, DC: The National Academies Press.

[52]Ben-Eliyahu, A., Rhodes, J.E., and Scales, P. (2014). The interest-driven pursuits of 15 year olds: "Sparks" and their association with caring relationships and developmental outcomes. *Applied Developmental Science, 18*(2), 76-89.

Engle, R.A., and Conant, F.C. (2002). Guiding principles for fostering productive disciplinary engagement: Explaining an emergent argument in a community of learners classroom. *Cognition and Instruction, 20*(4), 399-483.

[53]Hammer, D., Goldberg, F., and Fargason, S. (2012). Responsive teaching and the beginnings of energy in a third grade classroom. *Review of Science, Mathematics and ICT Education, 6*(1), 51-72.

[54]Hidi, S., and Renninger, K. (2006). The four-phase model of interest development. *Educational Psychologist, 41*(2), 111-127.

[55]Anderman, E.M., Eccles, J.S., Yoon, K.S., and Roeser, R. (2001). Learning to value mathematics and reading: Relations to mastery and performance-oriented instructional practices. *Contemporary Educational Psychology, 26*(1), 76-95.

[56]Yeager, D.S., and Dweck, C.S. (2012). Mindsets that promote resilience: When students believe that personal characteristics can be developed. *Educational Psychologist, 47*, 1-13.

Duckworth, A.L., Peterson, C., Matthews, M.D., and Kelly, D.R. (2007). Grit: Perseverance and passion for long-term goals. *Journal of Personality and Social Psychology, 92*, 1087-1101.

[57]Barton, A., and Roth, W. (2005). Rethinking scientific literacy. *Canadian Journal of Education 28*(3), 561-566.

National Research Council. (2009). *Learning Science in Informal Environments: People, Places, and Pursuits.* Committee on Learning Science in Informal Environments. P. Bell, B. Lewenstein, A.W. Shouse, and M.A. Feder (Eds.). Board on Science Education, Center for Education, Division of Behavioral and Social Sciences and Education. Washington, DC: The National Academies Press.

[58]Aikenhead, G. (2005). *Science Education for Everyday Life: Evidence-Based Practice.* Ways of Knowing in Science and Mathematics Series. New York: Columbia University, Teachers College Press.

Barron, B., and Bell, P. (forthcoming). Learning in informal and formal environments. In L. Corno and E. Anderman (Eds.), *Handbook of Educational Psychology: 3rd Edition*. Mahwah, NJ: Lawrence Erlbaum.

Grossman, J.M., and Porche, M.V. (2014). Perceived gender and racial/ethnic barriers to STEM success. *Urban Education, 49*(6), 698-727.

Mervis, J. (2015). A classroom experiment. *Science, 347*(6222), 602-605.

[59]Fusco, D., and Calabrese Barton, A. (2001). Representing student achievement in science. *Journal of Research in Science Teaching, 38*(3), 337-354.

Bang, M., Warren, B., Rosebery, A., and Medin, D. (2012). Desettling expectations in science education. *Human Development, 55*, 302-318.

Bell, P., Bricker, L.A., Tzou, C., Lee, T., and Van Horne, K. (2012). Engaging learners in scientific practices related to obtaining, evaluating, and communicating information. *The Science Teacher, 79*(8), 31-36.

Fusco, D. (2008). School vs. afterschool: A study of equity in supporting children's development. *Journal of Research in Childhood Education, 22*, 391-403.

Gutiérrez, K., and Vossoughi, S. (2010). "Lifting off the ground to return anew": Documenting and designing for equity and transformation through social design experiments. *Journal of Teacher Education, 61*(1-2), 100-117.

Rahm, J. (2008). Urban youths' hybrid positioning in science practices at the margin: A look inside a school-museum-scientist partnership project and an after-school science program. *Cultural Studies of Science Education, 3*, 97-121.

Mangels, J.A., Butterfield, B., Lamb, J., Good, C.D., and Dweck, C.S. (2006). Why do beliefs about intelligence influence learning success? A social-cognitive-neuroscience model. *Social, Cognitive, and Affective Neuroscience, 1*, 75-86.

[60]Emdin, C. (2011). Citizenship and social justice in urban science education. *International Journal of Qualitative Studies in Education, 24*(3), 285-301.

National Science Teachers Association. (2008). *Teaching Science to English-Language Learners: Building on Students' Strengths*. A Rosebery and B. Warren (Eds.). Arlington,VA: National Science Teachers Association Press.

[61]Vossoughi, S., Escudé, M., Kong, F., and Hooper, P. (2013). *Tinkering, Learning, and Equity in the After-School Setting*. Paper presented at the FabLearn Conference, Stanford University, Stanford, CA.

[62]Bang, M., and Medin, D. (2010). Cultural processes in science education: Supporting the navigation of multiple epistemologies. *Science Education, 94*(6), 1008-1026.

[63]Cobern, W.W., and Aikenhead, G.S. (1998). Cultural aspects of learning science. In B. Fraser and K. Tobin (Eds.), *International Handbook of Science Education* (Part One, pp. 39-52). Dordrecht, Netherlands: Kluwer Academic.

Falk, J.H. (2006). An identity-centered approach to understanding museum learning. *Curator, 49*(2), 151-166.

Wheaton, M., and Ash, D. (2008). Exploring middle school girls' ideas about science at a bilingual marine science camp. *Journal of Museum Education, 33*(2), 131-143.

[64]Porticella, N., Bonfield, S., DeFalco, T., Fumarolo, A., Garibay, C., Jolly, E., Huerta Migus, L., Pandya, R., Purcell, K., Rowden, J., Stevenson, F., and Switzer, A. (2013). *Promising Practices for Community Partnerships: A Call to Support More Inclusive Approaches to Public Participation in Scientific Research.* A report commissioned by the Association of Science-Technology Centers, Washington, DC. Available: http://www.birds.cornell.edu/citscitoolkit/promisingpractices/consensus-document/Consensus_Document.pdf/view [May 2015].

[65]Lave, J., and Wenger, E. (1990). *Situated Learning: Legitimate Peripheral Participation*. Cambridge, UK: Cambridge University Press.

Holland, D., Lachicotte, W., Skinner, D., and Cain, C. (1998). *Identity and Agency in Cultural Worlds.* Cambridge, MA: Harvard University Press.

[66]Larson, R., Eccles, J., and Gootman, J.A. (2004). Features of positive developmental settings. *Prevention Researcher, 11*(2), 8-13.

[67]Boaler, J. (2003). *Studying and Capturing the Case of the "Dance of Agency."* Presented at the 27th International Group for the Psychology of Mathematics Education Conference held jointly with the 25th PME-NA Conference, July 13-18, Honolulu, HI. Available: http://eric.ed.gov/?id=ED500873[May 2015].

Calabrese Barton, A., and Tan, E. (2010). "We be burnin!" Agency, identity, and science learning. *Journal of the Learning Sciences, 19*(2), 187-229.

Hull, G., and Greeno, J. (2006). Identity and agency in non-school and school worlds. In Z. Bekerman, N.C. Burbules, and D. Silberman-Keller (Eds.), *Learning in Places: The Informal Education Reader* (pp. 77-97). New York: Peter Lang.

[68]Nasir, N., and Hand, V. (2006). Exploring sociocultural perspectives on race, culture, and learning. *Review of Educational Research, 76*(4), 449-475.

[69]Luehmann, A.L. (2009). Students' perspectives of a science enrichment programme: Out-of-school inquiry as access. *International Journal of Science Education, 31*(13), 1831-1855.

[70]Azevedo, F.S. (2011). Lines of practice: A practice-centered theory of interest relationships. *Cognition and Instruction, 29*(2), 147-184.

Barron, B., Martin, C.K., Takeuchi, L., and Fithian, R. (2009). Parents as learning partners in the development of technological fluency. *International Journal of Learning and Media, 1*, 55-77.

Kafai, Y., and Peppler, K. (2011). Youth, technology, and DIY: Developing participatory competencies in creative media production. *Review of Research in Education, 35*(1), 89-119.

Vossoughi, S., and Bevan, B. (2014). *Making and Tinkering: A Review of the Literature.* Commissioned paper for Successful Out-of-School STEM Learning: A Consensus Study, Board on Science Education, June, National Research Council, Washington, DC. Available: http://sites.nationalacademies.org/cs/groups/dbassesite/documents/webpage/dbasse_089888.pdf [May 2015].

[71]Fusco, D. (2008). School vs. afterschool: A study of equity in supporting children's development. *Journal of Research in Childhood Education, 22*, 391-403.

[72]Azevedo, F.S. (2011). Lines of practice: A practice-centered theory of interest relationships. *Cognition and Instruction, 29*(2), 147-184.

Barron, B., and Darling-Hammond, L. (2008). Teaching for meaningful learning: A review of research on inquiry-based and cooperative learning. In L. Darling-Hammond (Ed.), *Powerful Learning: What We Know about Teaching for Understanding* (pp. 11-70). San Francisco: Jossey-Bass.

Barron, B., and Bell, P. (forthcoming). Learning in informal and formal environments. In L. Corno and E. Anderman (Eds.), *Handbook of Educational Psychology: 3rd Edition*. Mahwah, NJ: Lawrence Erlbaum.

[73]Calabrese Barton, A., and Tan, E. (2010). "We be burnin!" Agency, identity, and science learning. *Journal of the Learning Sciences, 19*(2), 187-229.

Rose, S., and Calabrese Barton, A. (2012). Should Great Lakes City build a new power plant? How youth navigate complex socioscientific issues. *Journal of Research in Science Teaching, 49*(5), 541-567.

[74]National Research Council. (2012). *A Framework for K-12 Science Education: Practices, Crosscutting Concepts, and Core Ideas*. Committee on Conceptual Framework for the New K-12 Science Education Standards. Board on Science Education, Division of Behavioral and Social Sciences and Education. Washington, DC: The National Academies Press.

National Research Council. (2009). *Learning Science in Informal Environments: People, Places, and Pursuits.* Committee on Learning Science in Informal Environments. P. Bell, B. Lewenstein, A.W. Shouse, and M.A. Feder (Eds.). Board on Science Education, Center for Education, Division of Behavioral and Social Sciences and Education. Washington, DC: The National Academies Press.

[75]Booker, A. (2010). Framing youth civic participation: Technical, pragmatic, and political learning. In L. Lin, H. Varenne, and E.W. Gordon (Eds.), *Educating Comprehensively: Varieties of Educational Experiences* (pp. 209-231). Lewiston, NY: Edwin Mellen Press.

Luehmann, A.L. (2007). Identity development as a lens to science teacher preparation. *Science Education, 91*(5), 822-839.

Mahoney, J.L., Harris, A.L., and Eccles, J.S. (2006). Organized activity participation, positive youth development, and the over-scheduling hypothesis. *Social Policy Report, 20*(4), 3-31. Available: http://files.eric.ed.gov/fulltext/ED521752.pdf [May 2015].

[76]Chavez, V., and Soep, E. (2005). Youth radio and the pedagogy of collegiality. *Harvard Educational Review, 47*(4), 409-434.

[77]O'Leary, E., and Soep, L. (2014). *Youth Radio's DO IT! Initiative Summative Evaluation: Final Project Report.* Available: http://informalscience.org/images/evaluation/2014-12-16_Youth%20Radio%20DO%20IT%20Final%20Report.pdf [February 2015].

[78]Barron, B. (2006). Interest and self-sustained learning as catalysts of development: A learning ecologies perspective. *Human Development, 49*, 193-224.

Barron, B., and Bell, P. (forthcoming). Learning in informal and formal environments. In L. Corno and E. Anderman (Eds.), *Handbook of Educational Psychology: 3rd Edition.* Mahwah, NJ: Lawrence Erlbaum.

Bevan, B., Bell, P., Stevens, R., and Razfar, A. (2012). *LOST Opportunities: Learning in Out-of-School Time.* Heidelberg, Germany: Springer Netherlands.

Bransford, J., Vye, N., Stevens, R., Kuhl, P., Schwartz, D., Bell, P., Meltzoff, A., Barron, B., Pea, R., Reeves, B., Roschelle, J., and Sabelli, N. (2006). Learning theories and education: Toward a decade of synergy. In P. Alexander and P. Winne (Eds.), *Handbook of Educational Psychology: Second Edition* (pp. 209-244). Mahwah, NJ: Lawrence Erlbaum.

[79]Mehus, S., Stevens, R., and Grigholm, L. (2010). *Interactional Arrangements for Learning about Science in Early Childhood: A Case Study Across Preschool and Home Contexts.* Paper presented at the 9th International Conference of the Learning Sciences, Chicago, IL.

[80]Bransford, J.D., Barron, B., Pea, R.D., Meltzoff, A., Kuhl, P., Bell, P., Stevens, R., Schwartz, D.L., Vye, N., Reeves, B., Roschelle, R., and Sabelli, N. (2005). Foundations and opportunities for an interdisciplinary science of learning. In R.K. Sawyer (Ed.), *The Cambridge Handbook of the Learning Sciences* (pp. 19-34). New York: Cambridge University Press.

Rahm, J. (2008). Urban youths' hybrid positioning in science practices at the margin: A look inside a school-museum-scientist partnership project and an after-school science program. *Cultural Studies of Science Education, 3*, 97-121.

Falk, J.H., and Needham, M. (2011). Measuring the impact of a science center on its community. *Journal of Research in Science Teaching, 48*(1), 1-12.

[81]Vossoughi, S., Escudé, M., Kong, F., and Hooper, P. (2013). *Tinkering, Learning, and Equity in the After-School Setting*. Paper presented at the FabLearn Conference, Stanford University, Stanford, CA.

[82]Azevedo, F.S., diSessa, A., and Sherin, B. (2012). An evolving framework for describing student engagement in classroom activities. *Journal of Mathematical Behavior, 31*, 270-289.

Falk, J.H., Dierking, L.D., Staus, N., Penuel, W., Wyld, J., and Bailey, D. (in press). Understanding and connecting youth STEM interest and participation across the community: The SYNERGIES project. *International Journal of Science Education, Part B.*

Falk, J.H., and Needham, M.D. (2013). Factors contributing to adult knowledge of science and technology. *Journal of Research in Science Teaching, 50*(4), 431-452.

[83]Ito, M., Gutiérrez, K., Livingstone, S., Penuel, W., Rhodes, J., Salen, K., Schor, J., Sefton-Green, J., and Watkins, S.C. (2012). *Connected Learning: An Agenda for Research and Design*. Chicago, IL: MacArthur Foundation.

Falk, J.H., Staus, N., Dierking, L.D., Wyld, J., Bailey, D., and Penuel, W. (2015). The SYNERGIES project: Preliminary results and insights from two years of longitudinal survey research. *Museology Quarterly, 29*(1), 15-21.

[84]Bevan, B., with Dillon, J., Hein, G.E., Macdonald, M., Michalchik, V., Miller, D., Root, D., Rudder, L., Xanthoudaki, M., and Yoon, S. (2010). *Making Science Matter: Collaborations Between Informal Science Education Organizations and Schools. A CAISE Inquiry Group Report.* Washington, DC: Center for Advancement of Informal Science Education.

Falk, J.H., Staus, N., Dierking, L.D., Wyld, J., Bailey, D., and Penuel, W. (2015). The SYNERGIES project: Preliminary results and insights from two years of longitudinal survey research. *Museology Quarterly, 29*(1), 15-21.

Stocklmayer, S.M., Rennie, L.J., and Gilbert, J.K. (2010). The roles of the formal and informal sectors in the provision of effective science education. *Studies in Science Education, 46*, 1-44.

Traphagen, K., and Traill, S. (2014). *How Cross-Sector Collaborations Are Advancing STEM Learning*. Palo Alto, CA: Noyce Foundation.

[85]Barron, B. (2006). Interest and self-sustained learning as catalysts of development: A learning ecology perspective. *Human Development, 49*, 193-224.

Bell, P., Tzou, C., Bricker, L.A., and Baines, A.D. (2012). Learning in diversities of structures of social practice: Accounting for how, why, and where people learn science. *Human Development, 55*, 269-284.

Falk, J.H., Staus, N., Dierking, L.D., Wyld, J., Bailey, D., and Penuel, W. (2015). The SYNERGIES project: Preliminary results and insights from two years of longitudinal survey research. *Museology Quarterly, 29*(1), 15-21.

[86]Falk, J.H., Dierking, L.D., Staus, N., Penuel, W., Wyld, J., and Bailey, D. (in press). Understanding and connecting youth STEM interest and participation across the community: The SYNERGIES project. *International Journal of Science Education, Part B.*

[87]Ito, M., Gutiérrez, K., Livingstone, S., Penuel, W., Rhodes, J., Salen, K., Schor, J., Sefton-Green, J., and Watkins, S.C. (2012). *Connected Learning: An Agenda for Research and Design.* Chicago, IL: MacArthur Foundation.

[88]National Research Council. (2009). *Learning Science in Informal Environments: People, Places, and Pursuits.* Committee on Learning Science in Informal Environments. P. Bell, B. Lewenstein, A.W. Shouse, and M.A. Feder (Eds.). Board on Science Education, Center for Education, Division of Behavioral and Social Sciences and Education. Washington, DC: The National Academies Press.

[89]Falk, J.H., and Dierking, L.D. (2010). The 95% solution: School is not where most Americans learn most of their science. *American Scientist, 98*, 486-493.

Volmert, A., Baran, M., Kendall-Taylor, N., and O'Neil, M. (2013). *You Have to Have the Basics Down Really Well: Mapping the Gaps Between Expert and Public Understanding of STEM Learning.* Washington, DC: Frameworks Institute. Available: http://www.frameworksinstitute.org/assets/files/PDF_STEM/STEMMTG10-18-13_proofedandformatted.pdf [February 2015].

[90]Lerner, R.M., and Lerner, J.V. (2013). *The Positive Development of Youth: Comprehensive Findings from the 4-H Study of Positive Youth Development.* Washington, DC: National 4-H Council.

National Research Council. (2009). *Learning Science in Informal Environments: People, Places, and Pursuits.* Committee on Learning Science in Informal Environments. P. Bell, B. Lewenstein, A.W. Shouse, and M.A. Feder (Eds.). Board on Science Education, Center for Education. Division of Behavioral and Social Sciences and Education. Washington, DC: The National Academies Press.

National Research Council and Institute of Medicine. (2002). *Community Programs to Promote Youth Development.* Committee on Community-Level Programs for Youth, J. Eccles, and J.A. Gootman (Eds.). Board on Children, Youth, and Families, Division of Behavioral and Social Sciences and Education. Washington, DC: National Academy Press.

[91]Bevan, B., Dillon, J., Hein, G.E., Macdonald, M., Michalchik, V., Miller, D., Root, D., Rudder, L., Xanthoudaki, M., and Yoon, S. (2010). *Making Science Matter: Collaborations Between Informal Science Education Organizations and Schools. A CAISE Inquiry Group Report.* Washington, DC: Center for Advancement of Informal Science Education.

Falk, J.H., Dierking, L.D., Osborne, J., Wenger, M., Dawson, E., and Wong, B. (2015). Analyzing science education in the UK: Taking a system-wide approach. *Science Education, 99*(1), 145-173.

Roth, W., and Van Eijck, M. (2010). Fullness of life as minimal unit: Science, technology, engineering, and mathematics (STEM) learning across the life span. *Science Education, 94*(6), 1027-1048.

[92]Volmert, A., Baran, M., Kendall-Taylor, N., and O'Neil, M. (2013). *You Have to Have the Basics Down Really Well: Mapping the Gaps Between Expert and Public Understanding of STEM Learning.* Washington, DC: Frameworks Institute. Available: http://www.frameworksinstitute.org/assets/files/PDF_STEM/STEMMTG10-18-13_proofedandformatted.pdf [February 2015].

[93]Bevan, B., Dillon, J., Hein, G.E., Macdonald, M., Michalchik, V., Miller, D., Root, D., Rudder, L., Xanthoudaki, M., and Yoon, S. (2010). *Making Science Matter: Collaborations Between Informal Science Education Organizations and Schools. A CAISE Inquiry Group Report.* Washington, DC: Center for Advancement of Informal Science Education.

Phillips, M., Finkelstein, D., and Wever-Frerichs, S. (2007). School site to museum floor: How informal science institutions work with schools. *International Journal of Science Education, 29*(12), 1489-1507.

Schatz, D., and Dierking, L.D. (1998). Systemic change in science education reform: Pacific Science Center. *Journal of Museum Education, 23*(2), 22-24.

[94]Dierking, L.D., Falk, J.H., Holland, G., Fisher, S., Schatz, D., and Wilke, L. (1997). *Collaborations: Critical Criteria for Success.* Washington, DC: Association of Science-Technology Centers.

[95]Thorhauge, S. (2014). Interface learning: New goals for museums and upper secondary school collaboration. *Dissertation Abstracts.* Available: http://pure.au.dk/portal/files/85300766/PhD_Interface_Learning_New_goals_for_museum_and_upper_secondary_school_collaboration.pdf [May 2015].

[96]Falk, J.H., Dierking, L.D., Staus, N., Penuel, W., Wyld, J., and Bailey, D. (in press). Understanding and connecting youth STEM interest and participation across the community: The SYNERGIES project. *International Journal of Science Education, Part B.*

[97]Falk, J.H., Staus, N., Dierking, L.D., Wyld, J., Bailey, D., and Penuel, W. (2015). The SYNERGIES project: Preliminary results and insights from two years of longitudinal survey research. *Museology Quarterly, 29*(1), 15-21.

[98]Banks, J.A., Au, K.H., Ball, A.F., Bell, P., Gordon, E. W., Gutiérrez, K.D., and Zhou, M. (2007). *Learning in and out of School in Diverse Environments: Life-long, Life-wide, Life-deep.* Seattle, WA: The LIFE Center, University of Washington, Stanford University, and SRI International and Center for Multicultural Education, University of Washington.

Barron, B., and Bell, P. (forthcoming). Learning in informal and formal environments. In L. Corno and E. Anderman (Eds.), *Handbook of Educational Psychology: 3rd Edition.* Mahwah, NJ: Lawrence Erlbaum.

[99]Farrin, L., and Mokros, J. (2012). Energy monitoring: Powerful connections between math, science, and community, *Science Scope, 36*(3), 23-28.

[100]Ito, M., Gutiérrez, K., Livingstone, S., Penuel, W., Rhodes, J., Salen, K., Schor, J., Sefton-Green, J., and Watkins, S.C. (2012). *Connected Learning: An Agenda for Research and Design.* Chicago, IL: MacArthur Foundation.

[101]Penuel, W.P., Lee, T., and Bevan, B. (2014). *Designing and Building Infrastructures to Support Equitable STEM Learning Across Settings.* Research + Practice Collaboratory Research Synthesis. San Francisco: The Exploratorium.

[102]Wellman, B., and Frank, K. (2001). Network capital in a multi-level world: Getting support from personal communities. In N. Lin, K. Cook, and R. Burt (Eds.), *Social Capital: Theory and Research* (pp. 233-273). Hawthorne, NY: Aldine de Gruyter.

[103]Stevens, R., O'Connor, K., Garrison, L., Jocuns, A., and Amos, D. (2008). Becoming an engineer: Toward a three-dimensional view of engineering learning. *Journal of Engineering Education, 97*(3), 355-368.

[104]Peter, N. (2009). Defining our terms: Professional development in out-of-school time. *Afterschool Matters, 8*(2), 34-41.

[105]Annie E. Casey Foundation. (2003). *The Unsolved Challenge of System Reform: The Condition of the Frontline Human Services Workforce.* Baltimore, MD: Author.

Cole, P. (2006). *Understanding the Afterschool Workforce: Opportunities and Challenges for an Emerging Profession.* Houston, TX: National AfterSchool Association.

[106]Peter, N. (2007). *Promising Practices in Out-of-School Time Professional Development.* Philadelphia, PA:

Out-of-School Time Resource Center, University of Pennsylvania.

[107]National Research Council and Institute of Medicine. (2002). *Community Programs to Promote Youth Development.* Committee on Community-Level Programs for Youth, J. Eccles and J.A. Gootman (Eds.). Board on Children, Youth, and Families, Division of Behavioral and Social Sciences and Education. Washington, DC: National Academy Press.

Peter, N. (2007). *Promising Practices in Out-of-School Time Professional Development.* Philadelphia, PA: Out-of-School Time Resource Center, University of Pennsylvania.

[108]For information on the distinctions and relationship among these three types of evaluation, see pp. 22-25 in Stufflebeam, D.L., and Shinkfield, A.J. (2007). *Evaluation Theory, Models, and Applications.* San Francisco: Jossey-Bass.

[109]Mezirow, J. (2000). *Learning as Transformation: Critical Perspectives on a Theory in Progress.* San Francisco: Jossey-Bass.

[110]Barron, B. (2014). *Formative Assessment for STEM Learning Ecosystems: Biographical Approaches as a Resource for Research and Practice.* Commissioned paper for Successful Out-of-School STEM Learning: A Consensus Study, Board on Science Education, June, National Research Council, Washington, DC. Available: http://sites.nationalacademies.org/cs/groups/dbassesite/documents/webpage/dbasse_089994.pdf [May 2015].

Falk, J.H., and Dierking, L.D. (2000). *Learning from Museums: Visitor Experiences and the Making of Meaning.* Walnut Creek, CA: AltaMira Press.

[111]Falk, J.H., and Dierking, L.D. (2000). *Learning from Museums: Visitor Experiences and the Making of Meaning.* Walnut Creek, CA: AltaMira Press.

Maltese, A.V., and Tai, R.H. (2010). Eyeballs in the fridge: Sources of early interest in science. *International Journal of Science Education, 32*(5), 669-685.

[112]Lemke, J.L., Lecusay, R., Cole, M., and Michalchik, V. (2015). *Documenting and Assessing Learning in Informal and Media-Rich Environments.* Cambridge, MA: MIT Press.

[113]Penuel, W.R., Fishman, B., Cheng, B.H., and Sabelli, N. (2011). Organizing research and development at the intersection of learning, implementation, and design. *Educational Researcher, 40*(7), 331-337.

Penuel, W.P., Lee, T., and Bevan, B. (2014). *Designing and Building Infrastructures to Support Equitable STEM Learning Across Settings.* Research + Practice Collaboratory Research Synthesis. San Francisco: The Exploratorium.

[114]Christensen, R., Knezek, G., and Tyler-Wood, T. (2015). A retrospective analysis of STEM career interest among mathematics and science academy students. *International Journal of Learning, Teaching, and Education Research, 10*(1), 45-58.

Garg, R., Kauppi, C., Urajnik, D., and Lewko, J. (2007). A longitudinal study of the effects of context and experience on the scientific career choices of Canadian adolescents. *Canadian Journal of Career Development, 9*(1), 15-24. Available: http://ceric.ca/cjcd/archives/v9-n1/article2.pdf [February 2015].

Jones, G., Taylor, A., and Forrester, J.H. (2011). Developing a scientist: A retrospective look. *International Journal of Science Education, 33*(12), 1653-1673.

[115]Michalchik, V., and Gallagher, L. (2010). Naturalizing assessment. *Curator, 53*(2), 209-219.

Shields, P. (2014). *The Palo Alto Convening on Assessment in Informal Science Settings: Synthesis Report.* Prepared for the Noyce Foundation. Menlo Park, CA: SRI International.

Gitomer, D. (2012). *Observational Methods for Assessment of Informal Science Learning and Education.* Commissioned paper for the Assessment of Informal and Afterschool Science Learning Workshop, Board on Science Education, National Research Council, June 11-12, Irvine, CA. Available: http://sites.nationalacademies.org/DBASSE/BOSE/DBASSE_080110 [February 2015].

Lemke, J.L., Lecusay, R., Cole, M., and Michalchik, V. (2015). *Documenting and Assessing Learning in Informal and Media-Rich Environments.* Cambridge, MA: MIT Press.

[116]Pierce, K.M., Auger, A., and Vandell, D.L. (2013). *Narrowing the Achievement Gap: Consistency and Intensity of Structured Activities During Elementary School.* Paper presented at the Society for Research in Child Development Biennial Meeting, April, Seattle, WA.

Vandell, D.L., Reisner, R.R., and Pierce, K.M. (2007). *Outcomes Linked to High-Quality Afterschool Programs: A Longitudinal Study of Promising Afterschool Programs.* Unpublished report prepared for the Charles Stewart Mott Foundation. Available: http://www.education.uci.edu/childcare/pdf/afterschool/PP%20Longitudinal%20Findings%20Final%20Report.pdf [February 2015].

[117]Zimmerman, H.T. (2012). Participation in science at home: Recognition work and learning in biology. *Journal of Research in Science Teaching, 49*(5), 597-630.

[118]Bathgate, M.E., Schunn, C.D., and Correnti, R. (2013). Children's motivation toward science across contexts, manner-of-interaction, and topic. *Science Education, 98*(2), 189-215.

Noam, G.G., and Shah, A.M. (2013). *Game Changers and the Assessment Predicament in Afterschool Science.* Belmont, MA: Program in Education, Afterschool, and Resiliency.

[119]Azevedo, F.S. (2011). Lines of practice: A practice-centered theory of interest relationships. *Cognition and Instruction, 29*(2), 147-184.

Barron, B. (2006). Interest and self-sustained learning as catalysts of development: A learning ecologies perspective. *Human Development, 49*, 193-224.

Herrenkohl, L.R., and Mertl, V. (2011). *How Students Come to Be, Know, and Do: A Case for a Broad View of Learning.* New York: Cambridge University Press.

Hull, G., and Greeno, J. (2006). Identity and agency in non-school and school worlds. In Z. Bekerman, N.C. Burbules, and D. Silberman-Keller (Eds.), *Learning in Places: The Informal Education Reader* (pp. 77-97). New York: Peter Lang.

Soep, E. (2006). Critique: Assessment and the production of learning. *The Teachers College Record, 108*(4), 748-777.

[120]Ellenbogen, K. (2014). *Summary of the CAISE Convening on Building Capacity for Evaluation in Informal Science, Technology, Engineering and Math (STEM) Education.* Washington, DC: Center for Advancement of Informal Science Education. Available: http://informalscience.org/research/ic-000- 000-010-034/ECB_Convening_Summary [February 2015].

Gitomer, D. (2012). *Observational Methods for Assessment of Informal Science Learning and Education.* Commissioned paper for the Assessment of Informal and Afterschool Science Learning Workshop, Board on Science Education, National Research Council, June 11-12, Irvine, CA. Available: http://sites.nationalacademies.org/DBASSE/BOSE/DBASSE_080110 [February 2015].

Krajcik, J. (2012). *Using the NRC Framework to Engage Students in Learning Science in Informal Environments.* Commissioned paper for the Assessment of Informal and Afterschool Science Learning Workshop, Board on Science Education, National Research Council, June 11-12, Irvine, CA. Available: http://sites.nationalacademies.org/DBASSE/BOSE/DBASSE_080110 [February 2015].

[121]Michalchik, V., and Gallagher, L. (2010). Naturalizing assessment. *Curator, 53*(2), 209-219.

[122]Noam, G., and Shah, A.M. (2013). *Game Changers and the Assessment Predicament in Afterschool Science.* Belmont, MA: Program in Education, Afterschool, and Resiliency.

[123]National Research Council. (2000). *How People Learn: Brain, Mind, Experience, and School—Expanded Edition.* Committee on Developments in the Science of Learning, J.D. Bransford, A.L. Brown, and R.R. Cocking (Eds.) and Committee on Learning Research and Education Practice, M.S. Donovan, J.D. Bransford, and J. Pellegrino (Eds.). Commission on Behavioral and Social Sciences and Education. Washington DC: National Academy Press.

National Research Council. (2009). *Learning Science in Informal Environments: People, Places, and Pursuits.* Committee on Learning Science in Informal Environments. P. Bell, B. Lewenstein, A.W. Shouse, and M.A. Feder (Eds.). Board on Science Education, Center for Education, Division of Behavioral and Social Sciences and Education. Washington, DC: The National Academies Press.

[124]National Research Council. (2009). *Learning Science in Informal Environments: People, Places, and Pursuits.* Committee on Learning Science in Informal Environments. P. Bell, B. Lewenstein, A.W. Shouse, and M.A. Feder (Eds.). Board on Science Education, Center for Education, Division of Behavioral and Social Sciences and Education. Washington, DC: The National Academies Press.

[125]National Research Council. (2009). *Learning Science in Informal Environments: People, Places, and Pursuits.* Committee on Learning Science in Informal Environments. P. Bell, B. Lewenstein, A.W. Shouse, and M.A. Feder (Eds.). Board on Science Education, Center for Education, Division of Behavioral and Social Sciences and Education. Washington, DC: The National Academies Press.

National Research Council and Institute of Medicine. (2002). *Community Programs to Promote Youth Development.* Committee on Community-Level Programs for Youth. J. Eccles and J.A. Gootman (Eds.). Board on Children, Youth, and Families, Division of Behavioral and Social Sciences and Education. Washington, DC: National Academy Press.

APPENDIX A
Successful Out-of-School STEM Learning Summit Agenda

Out-of-School STEM Learning: A National Summit
June 3 and 4

National Academy of Sciences (NAS)
2100 Constitution Ave
Washington DC

Day 1: NAS Auditorium
Day 2: Lecture Room

June 3: NAS Auditorium

8:00 am **Networking** (coffee and light refreshments)
Poster Set-Up for Lunch-Time Session

8:30 am **Welcome and Overview**
Speakers
Martin Storksdieck, Director, Board on Science Education
Eric Jolly, President, Science Museum of Minnesota, Committee Chair
Joan Ferrini-Mundy, Assistant Director, Directorate for Education and Human Resources, National Science Foundation (Invited)
Dennis Schatz, National Science Foundation

9:15 am **What and Where Is STEM Learning Occurring**
Moderator
Nancy Peter, Out-of-School Time Resource Center, Committee Member
Speakers
Lynn S. Liben, Pennsylvania State University, Committee Member
John Falk, Oregon State University, Committee Member
Respondents
Ron Ottinger, Executive Director, Noyce Foundation
Andrea Ingram, Vice President of Education and Guest Services, Chicago Museum of Science and Industry
Ellen Gannet, Director, National Institute on Out-of-School Time
Ellen Lettvin, Robert Noyce Fellow in Informal STEM Learning, U.S. Department of Education

10:45 am **Audience Reflection on Workshop Goals and Out-of-School STEM Learning**
Moderator
Michael Feder, Study Director

11:05 am **What Is Success**
Moderator
Milbrey McLaughlin, Stanford University, Committee Member
Speakers
Jacque Eccles, University of California, Irvine, Committee Member
Karen Pittman, President and CEO, Forum for Youth Investment
Anita Krishnamurthi, Vice President, STEM Policy, Afterschool Alliance

12:35 pm **Poster Session** (lunch served)

1:30 pm **Characteristics of Successful Out-of-School STEM Learning Efforts**
Moderator
Vera Michalchik, Stanford University, Committee Member
Speakers
Bronwyn Bevan, Exploratorium, Committee Member
Emilyn Green, Executive Director, Community Science Workshop Network
Nathaniel Kendall-Taylor, Senior Researcher, Frameworks Institute

2:45 pm **Expanding Access to STEM Learning**
Moderator
Cary Sneider, Portland State University Committee Member
Speakers
Sue Allen, Director of Research at Maine Mathematics and Science Alliance
Saskia Traill, Vice President for Policy and Research, The After-School Corporation
Maria Cabrera, Community Relations, Museum of Science, Boston

3:55 pm **Break**

4:05 pm **Audience Reflection on Success**
Moderator
Michael Feder, Study Director

4:30 pm **Day 1 Themes and Take-a-Ways**
Moderator
Eric Jolly, Science Museum of Minnesota, Committee Chair
Discussants
Committee members

5:00 pm **Speed Networking** (optional, light refreshments)

5:30 pm **Adjourn**

June 4: Lecture Room

8:00 am	**Poster Session and Networking** (coffee and light refreshments)
8:30 am	**Welcome and Overview** Speaker *Eric Jolly, Science Museum of Minnesota, Committee Chair*
8:45 am	**Understanding and Assessing Success** Moderator *Bronwyn Bevan, Exploratorium, Committee Member* Speakers *David Hammer, Tufts University* *Phil Bell, Washington University* *Brigid Barron, Stanford University*
10:15 am	**Out-of-School STEM Learning Exemplars** Breakout Session 1: Local and National Youth-Serving Programs (Lecture Hall) Moderator *Nancy Peter, Out-of-School Time Resource Center, Committee Member* Speakers *Chad Ripberger, Rutgers University, 4-H STEM* *Jason Lee, DAPCEP* *Jill Walahoski, Nebraska University, Committee Member* Breakout Session 2: Youth-Driven STEM Experiences (Room 118) Moderator *Jane Buikstra, Arizona State University, Committee Member* Speakers *Rick Bonney, Cornell University* *Natalie Rusk, MIT Media Lab* *Gail Breslow, Computer Clubhouse Network* *Bronwyn Bevan, Exploratorium, Committee Member* Breakout Session 3: STEM Programs Managed by Museums, Science Centers, Etc. (West Court) Moderator Eric Jolly, *Science Museum of Minnesota, Committee Chair* Speakers *Kirsten Ellenbogen, Great Lakes Science Museum* *Dale McCreedy, Franklin Institute* *Bernadette Chi, Lawrence Hall of Science*

Breakout Session 4: After-school, Informal, and School Collaborations (Room 250)
Moderator
Maya Garcia, DC Office of the State Superintendent of Education, Committee Member
Speakers
James Short, American Museum of Natural of History
Debbie Zipes, Indiana Afterschool Network
Minda Borun, Franklin Institute

11:45 pm **Poster Session** (lunch served)

12:45 pm **Systems for Successful Out-of-School STEM Learning**
Moderator
Cary Sneider, Portland State University, Committee Member
Panelists
Michael Funk, After-School Division, California State Department of Education
Kevin Crowley, University of Pittsburgh
Linda Kekelis, Techbridge

2:15 pm **Policy Maker Reflections on Out-of-School STEM Learning**
Moderator
Eric Jolly, Science Museum of Minnesota, Committee Chair
Panelists
Tom Payzant, Harvard University, and former Superintendent of Boston Public Schools
James Geringer, Director of Policy at Environmental Systems Research Institute, and former Governor of Wyoming
Mary Lord, President-Elect, National Association of the State Boards of Education, and the American Society for Engineering Education

3:15 pm **Break**

3:30 pm **Workshop Themes and Lessons**
Moderator
Michael Feder, Study Director

4:30 pm **Final Thoughts**
Moderator
Eric Jolly, Science Museum of Minnesota, Committee Chair
Panelists
Committee Members

APPENDIX B
Papers Commissioned for the Study

The committee commissioned six papers to synthesize the research and evaluation studies that relate to our charge:*

- *Formative Assessment for STEM Learning Ecosystems: Biographical Approaches as a Resource for Research and Practice* by Brigid Barron
- *Citizen Science and Youth Education* by Rick Bonney, Tina B. Phillips, Jody Enck, Jennifer Shirk, and Nancy Trautmann
- *Evidence & Impact: Museum-Managed STEM Programs in Out-of-School Settings* by Bernadette Chi, Rena Dorph, and Leah Reisman
- *Children Doing Science: Essential Idiosyncrasy and the Challenges of Assessment* by David Hammer and Jennifer Radoff
- *Broadening Access to STEM Learning Through Out-of-School Learning Environments* by Laura Huerta Migus
- *Making and Tinkering: A Review of the Literature* by Shirin Vossoughi and Bronwyn Bevan

---

*The background papers are available at http://sites.nationalacademies.org/DBASSE/BOSE/CurrentProjects/DBASSE_086842 [May 2015].

APPENDIX C
Board on Science Education

ADAM GAMORAN (*Chair*), William T Grant Foundation (president), New York

GEORGE BOGGS, Palomar College, San Marcos, CA (emeritus)

MELANIE COOPER, Department of Chemistry, Michigan State University

RODOLFO DIRZO, Department of Biology, Stanford University

JACQUELYNNE ECCLES, Department of Psychology, University of Michigan

JOSEPH FRANCISCO, Department of Chemistry, Purdue University

MARGARET A. HONEY, New York Hall of Science, New York

SUSAN KIEFFER, Department of Geology, University of Illinois, Urbana

MATTHEW KREHBIEL, Kansas State Department of Education, Topeka

MICHAEL LACH, Urban Education Institute, University of Chicago

LYNN S. LIBEN, Department of Psychology, Pennsylvania State University

BRIAN REISER, School of Education and Social Policy, Northwestern University

MARSHALL "MIKE" SMITH, Carnegie Foundation for the Advancement of Teaching, Stanford, CA

ROBERTA TANNER, Physics Teacher (retired), Thompson School District, Loveland, CO

SUZANNE WILSON, Neag School of Education, University of Connecticut

YU XIE, Department of Sociology, University of Michigan

HEIDI SCHWEINGRUBER, *Director*

MICHAEL FEDER, *Senior Program Officer*

MARGARET HILTON, *Senior Program Officer*

MATT LAMMERS, *Program Coordinator*

KELLY ARRINGTON, *Senior Program Assistant*

JOANNA ROBERTS, *Program Assistant*

APPENDIX D
Acknowledgments

This report is made possible by the important contributions of the National Research Council (NRC) staff, the study committee, and many other experts. First, we acknowledge the sponsorship of the National Science Foundation (NSF). We particularly thank NSF program officers Denis Schatz, Al DeSena, and Julie Johnson.

This report was informed by a National Summit on Successful Out-of-School STEM L-earning on June 3-4, 2014. Organized by the study committee, the summit was held in Washington, D.C. On behalf of the committee, I would like to thank the people who made presentations at the summit: for videos of the presentations, and slides used by the presenters, see http://sites.nationalacademies.org/DBASSE/BOSE/DBASSE_086989 [February 2015]. The summit agenda is in Appendix A, and the list of commissioned papers presented at the summit is in Appendix B.

This report has been reviewed in draft form by individuals chosen for their diverse perspectives and technical expertise, in accordance with procedures approved by the Report Review Committee of NRC. The purpose of this independent review is to provide candid and critical comments that will assist the institution in making its published report as sound as possible and to ensure that the report meets institutional standards for objectivity, evidence, and responsiveness to the charge. The review comments and draft manuscript remain confidential to protect the integrity of the process.

I thank the following individuals for their review of this report: Sue Allen, Research, Maine Mathematics and Science Alliance, Augusta, ME; James Bell, Center for Advancement of Informal Science Education, Association of Science-Technology Centers, Washington, D.C.; William B. Bridges, Department of Engineering Emeritus, California Institute of Technology; Ilan Chabay, Institute for Advanced Sustainability Studies, Potsdam, Germany; Ellen S. Gannett, National Institute on Out-of-School Time, Wellesley Centers for Women; Richard M. Lerner, Institute for Applied Research in Youth Development, Tufts University; Lester L. Lyles, Independent Consultant, The Lyles Group, Vienna VA; Dale McCreedy, Gender, Adult Learning and Community Engagement, The Franklin Institute, Philadelphia, PA; David Pines, Department of Physics, University of Illinois at Urbana–Champaign; and Robert M. West, Informal Learning Experiences, Denver, CO.

Although the reviewers listed above provided many constructive comments and suggestions, they were not asked to endorse the content of the report nor did they see the final draft of the report before its release. The review of this report was overseen by Eugenie C. Scott, National Center for Science Education, and Stephen R. Berry, University of Chicago. Appointed by NRC, they were responsible for making certain that an independent examination of this report was carried out in accordance with institutional procedures and that all review comments were carefully considered. Responsibility for the final content of this report rests entirely with the committee and the institution.

Thanks are also due to the project staff and staff of the Division of Behavioral and Social Sciences and Education (DBASSE). Joanna Roberts managed the study's logistical and administrative needs, making sure meetings and workshops ran efficiently and smoothly. Eugenia Grohman of the DBASSE staff substantially improved the readability of the report. We are also grateful to Argenta Price (Christine Mirzayan Science and Technology Fellow) for her contribution to the study. Kirsten Sampson Snyder of the DBASSE staff expertly guided us through the NRC review process, and Yvonne Wise of the DBASSE staff oversaw the production of the report. Most importantly, Michael Feder, NRC Board on Science Education, directed the study and played a key role in the report drafting process.

Eric Jolly, *Chair*
Committee on Successful Out-of-School STEM Learning

PHOTO CREDITS

Front cover: (top row, left to right) iStock image #36932556, ©Pamela Moore; iStock image #40808888, ©Imgorthand; iStock image #25906670, ©Goodluz; iStock image #67320243, ©Marilyn Nieves (second row, left to right) iStock image #43832016, ©SolStock; iStock image #49190352, ©PeopleImages; iStock image #24899236, ©Pamela Moore; iStock image #37936462, ©JillianSuzanne (third row, left to right) iStock image #57206936, ©Marilyn Nieves; iStock image #41901540, ©energyy; iStock image #48632996, ©franckreporter; iStock image #42181580, ©Steve Debenport (fourth row, left to right) iStock image #61088534, ©vitranc; iStock image #45592470, ©Photo_Concept; iStock image #52287768, ©PamelaJoeMcFarlane; iStock image #52169092, ©Prasit Rodphan

page vi: iStock image #47618648, ©stevecoleimages; Page 1: iStock image #40808888, ©Imgorthand; Page 8: iStock image #57206936, ©Marilyn Nieves; Page 10: iStock image #45592470, ©Photo_Concept; Page 14: iStock image #36932556, ©Pamela Moore; Page 17: iStock image #24899236, ©Pamela Moore; Page 18: iStock image #41901540, ©energyy; Page 23: iStock image #52169092, ©Prasit Rodphan; Page 27: iStock image #42181580, ©Steve Debenport; Page 30: iStock image #43832016, ©SolStock; Page 31: iStock image #61088534, ©vitranc; Page 32: iStock image #48632996, ©franckreporter; Page 36: iStock image #25906670, ©Goodluz; Page 38: iStock image #49190352, ©PeopleImages; Page 41: iStock Image #67320243, ©Marilyn Nieves; Page 42: iStock image #52287768, ©PamelaJoeMcFarlane; Page 45: iStock image #37936462, ©JillianSuzanne